A Guide to the Systems of Provision Approach

"Bayliss and Fine's book on the theory and application of the "systems of provision" approach is a social science tour-de-force, reconnecting production and consumption well beyond the over simplistic supply and demand of neoclassical economics. In our time of climate crisis, the insights arising from systems of provision are not just illuminating: they may prove crucial to spur the necessary transformation of our economies."

—Julia Steinberger, *Professor of Political Ecology, University of Leeds.*

"The climate crisis has now arrived, and it will require radical changes in consumption patterns, especially in a context of steady erosion of public goods and services. These changes in consumption cannot be understood with mainstream models of supply and demand; instead, they involve complex interactions all along supply chains, whilst practices of consumption involve not only market signals but cultural understandings. In this path breaking book, Kate Bayliss and Ben Fine draw from decades of rich empirical case studies to synthesize lessons into a comprehensive method for understanding consumption. Their concept of 'systems of provision' breaks new theoretical ground, offering social scientists and policymakers a way forward for comprehensively understanding and better shaping systems of provision in contemporary capitalism. Social policy experts and scholars in particular now have a powerful analytical key to new insights into who wins and who loses from the (re-)commodification of the sphere of social reproduction."

—Lena Lavinas, *Professor of Welfare Economics, Institute of Economics, Federal University of Rio de Janeiro.*

"Consumption is too important a topic to be left to economists, but can the other social sciences come up with something as all-embracing and coherent? The Systems of Provision approach claims to provide this alternative framework. Consumption in all its complexity can only be understood in a disaggregated way: the energy system differs from the fashion system, the food system from transport, the hospitality system from health care. This book provides a rich application of these ideas and a guide to policy making and activism, all the more important as runaway consumption in the rich world threatens all our futures."

—Ian Gough, Visiting Professor, *London School of Economics and Emeritus Professor, University of Bath.*

"Who consumes how much of what, and why? This is not really about "individual preferences", but rather is critically determined by how goods and services are provided, and how integrated chains of provisioning are intertwined with material cultures. This important book will revolutionise how you think about the consumption of items as disparate as food, consumer durables, or education."

—Jayati Ghosh, *Professor of Economics, Jawaharlal Nehru University, New Delhi.*

"This is a must-read for anyone interested in consumption studies. Empirically based and theoretically rigorous, it provides a comprehensive but accessible account of the SOP approach—which has established itself as the most truly interdisciplinary approach to understanding consumption in all its diverse and complex contexts."

—Desmond McNeill, *Centre for Development and the Environment, University of Oslo.*

"Bayliss and Fine's systems of provision (SoP) approach offers a useful grand framework for the social analysis of consumption. They go beyond what is consumed and provided to recognize various rationales for consumption. These include norms, incorporating socially and culturally determined propensities to consume, and the spectrum of provisioning, ranging from public goods to private suppliers. As their book includes both theorization and wide-ranging case studies, the approach offers alternative economic theory and analysis of consumption, besides informing related analyses."

—Jomo Kwame Sundaram, *Visiting Senior Fellow at Khazanah Research Institute, Visiting Fellow at the Initiative for Policy Dialogue, Columbia University, and Adjunct Professor at the International Islamic University in Malaysia.*

Kate Bayliss • Ben Fine

A Guide to the Systems of Provision Approach

Who Gets What, How and Why

palgrave
macmillan

Kate Bayliss
Sustainability Research Institute
University of Leeds
Leeds, UK

Department of Economics
SOAS, University of London
London, UK

Ben Fine
Department of Economics
SOAS, University of London
London, UK

Wits School of Governance
University of Witwatersrand
Johannesburg, South Africa

ISBN 978-3-030-54142-2 ISBN 978-3-030-54143-9 (eBook)
https://doi.org/10.1007/978-3-030-54143-9

This Palgrave Macmillan imprint is published by the registered company Springer Nature Switzerland AG.
The registered company address is: Gewerbestrasse 11, 6330 Cham, Switzerland

Preface

We are often asked in casual conversation and even more in scholarly exchanges to sum up in a sound bite or three what is a system of provision and what is the systems of provision approach. This foreword offers some sort of response, but it is more of a bite than a light snack, giving some sort of flavour of the heavier meal or, more exactly, comprehensive diet and cuisine, that follows in this book. So, here goes.

"Systems of provision (SoP)" is an approach to the study of consumption that synthesises core elements from across the social sciences in a coherent and manageable framework. This is important because the question of "who has what, how and why" is central to human existence. Fundamental aspects of our everyday lives, such as access to food, water, shelter, schools and health services, hinge on such issues. Pressing global concerns such as climate change, inequality and public health drill down to these core themes which are all the more pressing in light of the emerging global coronavirus pandemic.

Across the social sciences different disciplines have applied their own slant to try to explain what it is that determines what we consume. This is particularly significant for much of modern economics where consumption outcomes continue to be seen to revolve around the hypothetical actions of atomistic, utility-maximizing individuals, operating in the context of more or less perfectly functioning markets.

Other disciplines approach the causes of consumption differently, more closely linked to what it is that is being consumed, who is consuming and in light of culture and context. For example, anthropology focuses on the associations and meanings of objects and the ways that societies interact with these. Sociology sees consumption as related to identity and social order and is concerned with the role played by social class, race and income levels as a marker of stratification. Psychology plays a role in modern marketing studies which are interested in how shopping habits can be influenced. Scholars of feminist theory work across social science disciplines, for example on how consumption links to the gender division of labour or how advertising shapes notions of gender.

Clearly, then, the range of factors that determine who has what is huge. And will vary depending on the person, the object and the context. How then can we capture the range and complexity of these multifarious factors? The SoP approach was devised in the 1990s in response to the failings of the social sciences individually and collectively to provide a coherent understanding of consumption. The approach has evolved into a structured framework that draws on the scope of social sciences with the notable exception of mainstream economics, of which the SoP approach has been heavily critical.

The essence of the SoP approach is that consumption outcomes depend on the system by which a good or service is provided. This in turn is shaped by the nature of the good itself and the context in which both production and consumption, and the connections between them, are situated. The approach highlights the importance of the specific context. Put simply, the factors which shape consumption of peanut butter will be different from those for education or motor vehicles which will be different again across countries and different today from a century ago. In contrast with mainstream economics, and in line with other social sciences, consuming is seen as a social rather than an individual practice oriented around what are often well-established patterns of consumption, or consumption norms. At the simplest level (sound bite even), and as the starting point for much complexity, the SoP approach is driven by the notion, prominent in popular understanding, that there is an energy system and this is different from the health system that is, in turn, different from the transport system, the fashion system, the food systems and so on, and each needs to be acknowledged as such on its own merits or flaws.

But how to understand these different systems and the differences between them? The SoP approach synthesises the complex drivers of consumption into five broad themes, which are discussed below. It is important to note that these themes are interrelated and the boundaries between them are blurred. This should not be considered as a methodological framework, so much as guide to core issues to bear in mind.

Agents—SoPs are determined by the participants in the system. This includes (individual) consumers themselves but equally producers and other bodies that might be involved such as trade unions, consumer groups and the state. There will be sub-groups within these so that producers might include financiers, shareholders, management and workers. Consumers might be differentiated according to different strata depending on what is being analysed. The state will also have different functions, possibly as regulator, policy maker or even provider. SoP analysis needs to map the agents involved in the system and to consider their role, including their motives and rationales for engaging within or with the SoP. These factors will inevitably be contested, with conflict across different categories of agents—see *relations* below—rather than be harmoniously integrated and coordinated through some amorphous market (or other institutions).

Structures—The SoP will be situated within a wider context which will shape the way that goods are produced and consumed. Structures take different forms such as organizational, institutional and social, formal or informal. There may be legal structures, for example, that shape the SoP. Examples of social structures that would impact on a SoP might be along the lines of income levels, race and/or gender. Structures will intersect. In the housing sector, for example, the institutional structures such as different forms of housing tenure (owner-occupied, private rental, social housing) combine with social structures to lead to social outcomes which might take the form of inequalities in levels of, and access to, housing. SoP analysis can consider how consumption might vary across different structural components and what the impact of these might be. SoP studies may also be concerned with the origins of such structures, which could be connected with other SoP themes such as ***material culture***; see below for some light coverage of some examples considered in depth in the main text of this book.

Processes—These can relate to specific activities in the SoP such as how labour is employed, how advertising is used. Other more abstract processes will also shape the SoP such as processes of globalisation, privatisation and financialisation. Processes link to the other SoP elements, for example, privatisation may change the nature of the *agents* and/or the *structures* of provisioning as the state/market *relation* is reconfigured.

Relations—On one level, the SoP analysis will be concerned with the relations between the *agents*. The priorities of the different groups may be conflicting which means that the way the SoP operates will be contingent on who exercises power and how. At a broader level, relations in society may be shaped by its *structures* such as social class, gender and race. The SoP approach in part derives from the observation that consumption tends to happen in clustered patterns rather than on an individual basis, and corresponding consumption norms are often segregated along these structural formations. SoP analysis, therefore, will be concerned with the role played by social norms and the way that agents, relations and structures shape these.

Material cultures—Each SoP will also be characterized by different meanings for the *agents* who are involved with it in whatever way, including producers and consumers as well as citizens, policymakers, the media and scholars. Systems are reproduced, or not, by prevailing narratives. The SoP approach seeks to unpack what these narratives are and who or what leads to them being created, sustained or contested and transformed. The *agents* in the SoP will have different associations with it. For example, when it comes to infrastructure (such as roads and rail tracks, water pipes and electricity lines), this is seen differently by those who build and finance it, governments who sponsor it, people who use it and people who oppose its construction. And these meanings will change over time as, for example, a new airport runway, which might once have been seen as the height of modernity, can now be considered an environmental disaster. Material cultures are often so deeply ingrained that they can be difficult to identify. Corresponding meanings can also be malleable and fleeting, hence the heavy presence of advertising, and control of, and influence over, the material culture of consumption plays an important role in the shaping of the SoP more broadly—whether it be the construction of owner-occupation as the ideal form of housing or the private motor car as that for transport.

Clearly it will be beyond the scope of most research projects and activist literature to capture each and every element of the SoP. The researcher will need to narrow down to the specific aspects that are of interest. When it comes to methodology, research using the SoP approach tends to be qualitative, oriented around specific case studies but that is not to say that more quantitative methods could not be used. An ever-present challenge in conducting SoP research is that it is not clear from the outset where one SoP ends and another begins as ultimately everything connects to everything else. There will be a SoP for food in general, but also for dairy and even for cheese in particular. Boundaries cannot be set in advance but emerge inductively as part of the research process.

The approach provides a way to unpack the opaque workings of who gets what and how, as well as why. Equally significant is who is excluded from consumption and with what effects. This book contains numerous examples of research studies that have drawn on the SoP approach. These range in scale and scope from health services in England to international supply chains for orange juice. The approach has been shown to be durable and versatile, with varieties of applications. Some studies have followed the approach in detail, examining the structures, processes, agents and relations which underpin provisioning as well as the material cultures that shape consumption decisions. Others have taken the SoP approach as one of a group of approaches on which to draw when considering consumption. Still others just use some core elements of the approach selectively, as for example, when the focus is on material cultures.

Applying this framework draws attention to the systemic factors that determine who gets what, not just in terms of end consumption but also how value is distributed in the chain of provisioning. Given the importance of specificity, each study raises different issues but these are often are linked by systemic processes and conceptualisations, such as neoliberalism and globalisation. Two examples are briefly highlighted as tasters for what is to come.

In *housing* in the UK the superficial explanation of who has what is one of a shortage of housing supply pushing up prices thereby lowering affordability. The policy response is to increase housebuilding. SoP analyses, in contrast, unpack the agents and the history of housing, linking the current structure to the past dismantling of social housing and to a financial system that has instilled a speculative nature in housing production and ownership. In addition, SoP studies emphasise the changing material culture of housing where home ownership has since the 1980s been represented as innately preferable to other types of tenure, and housing has been constructed as a financial asset. This cultural shift links to a broader neoliberal understanding where individuals bear responsibility for their own welfare. Through the SoP lens, the policy response of increasing housing construction risks increasing speculative activity in the housing market, and ownership ending up in the hands of financial investors. Instead, to alleviate the increasing vulnerability of low-income households, attention is needed to social housing and to alternative systems of finance which are oriented around social provision.

When it comes to *food*, the world faces major health risks due to over consumption. The incidence of obesity has doubled in the past three decades. The mainstream approach has been to address the consumer, as if the cause was a sudden lack of knowledge and will power. Public health initiatives to encourage healthy eating are commonplace but have had little success. SoP takes a different approach. First is to look at the production of food. The food SoP rewards ever-expanding food production, with profits to be made from dysfunctional practices such as land grabbing and speculation in agricultural commodity prices. Expanding production needs to be consumed, so providers have a compulsion to create a need to eat that goes far beyond physiological need. Producers make use of advertising as well as changing distribution systems, from out-of-town supermarkets to ubiquitous snack retail outlets to push food sales. The second part of the SoP approach to understanding food consumption is to examine the way in which the consumer engages with this provisioning system, facing pressure both to eat and to diet. The SoP approach is concerned with the cultures that have grown up around food, viewing the epidemic of obesity emerging as in part due to the systemic imperative to accommodate profitable supply.

These and other examples in this book, such as the SoPs for water and "fast" fashion, demonstrate that the act of consumption connects the consumer, usually unwittingly, with the underlying social relations of production. These can be extensive. In fashion for example, the range of agents in the SoP reaches across continents from agricultural workers and machinists usually from the Global South to retailers, designers, models and advertisers often in the North. In water in England, the apparently mundane act of turning on the tap, ostensibly unchanged for decades, is now in some cases to connect to a world of private equity and off shore finance. These social relations can be exploitative and this comes to light in SoP studies. Clothes, for example, are constructed through a web of relationships which draw upon and reproduce positions of power and dependency emerging from decades, even centuries, of uneven development.

The mainstream narrative of markets and individual choice leads to consumer-oriented interventions to promote change. But rather than alleviating the adverse social effects these can further embed inequality. Campaigns around ethical consumption such as Fairtrade, for example, may bring some slight improvements to the conditions of wage labour but the underlying conditions are unchanged. Individual adjustments do not lead to widespread change. In road transport, SoP analysis has shown that persistent car use results from the intersection of a number of dynamics including the industrial structure of the automotive industry, the nature of the built environment and cultures of car consumption. As with food, scale economies in car production create an inherent tendency to overproduction which then needs to be consumed. A culture of economic growth presents greater car consumption as inherently a good thing.

Thus, the SoP approach can draw attention to the extensive and complex channels which underpin specific social, environmental and economic outcomes, and can therefore be helpful in identifying "leverage points" for interventions and pressure for change. But it is also instructive in pinpointing the challenges involved. Transformational change is likely hindered by a reluctance to relinquish home comforts and is also impeded by deeply embedded narratives and structures that lock in the status quo.

SoPs themselves as well as their core elements are often under the radar. The studies cited in this book show that the consumer is often distanced from the social relations that underpin provisioning both in policy and mainstream narratives. Such distancings or, more exactly, concealments serve to focus attention on the commodity and on the capacities of the consuming individual. With a focus on individuals and markets, the underlying systems and the cultures by which these are sustained are rendered almost invisible.

The SoP approach aims to shine a light on these opaque issues. It provides a systematic framework to address the extensive and diverse factors that shape provisioning outcomes. When the SoP approach lifts the lid on the production system, and its connections to consumption, the ways that agents are competing to capture value in the system are revealed. Outcomes emerge from contested social relations embedded in longstanding structures and processes. The approach goes beyond the simple tracking of value chains and agents, to consider how these interact and how the state is involved in promoting specific outcomes.

The approach has proven to be remarkably durable, offering an open, flexible, if uncompromisingly grounded, analytical framing. It provides insights at the grand scale, for example, relating to inequality, the environment, the pandemic and the crisis of dietary diseases as well as the everyday access to health, water, housing and so on. As such, the SoP approach has much to offer for the challenges ahead in the worlds of scholarship, popular discourse, policymaking and activism.

London, UK Kate Bayliss
London, UK Ben Fine

Acknowledgements

We would like to thank all of those with whom we have engaged over the years in contributing directly and indirectly, as well as critically, to the systems of provision (SoP) approach. Particularly gratifying have been those who have deployed the approach without necessarily having a personal connection. All of these contributions to forming and expressing our thinking have been invaluable, and this is especially so of co-authors and fellow researchers, many but not all of whom can be gleaned from our lists of references.

The approach would not have advanced over the years without support and faith from a number of collaborators and flag bearers, particularly at the University of Leeds, notably Andrew Brown and Julia Steinberger.

We are grateful to the following recent financial sponsors of our research:

The Leverhulme Trust Research Leadership Award for the research project 'Living Well Within Limits' (LiLi) which provided support during the drafting of this book.

The European Union's 7th Framework Programme for research, technological development and demonstration grant for the FESSUD project (Financialisation, Economy, Society and Sustainable Development) which funded a number of case studies using the systems of provision approach (see Chaps. 1 and 6).

And, during the 1990s, the UK's, Economic and Social Research Council (ESRC) funded two research projects, one on the ownership of consumer durables and female labour market participation, and the other on patterns of food consumption. Together these were vital in getting the SoP approach going (see Chaps. 1, 5 and 6).

Contents

Acronyms and Abbreviations

ANC	African National Congress
ESRC	Economic and Social Research Council
EU	European Union
EW	England and Wales
FESSUD	Financialisation, Economy, Society and Sustainable Development
FTEPR	Fair Trade, Employment and Poverty Reduction
GCC	Global Commodity Chains
GFC	Global Financial Crisis
GPN	Global Production Networks
GVC	Global Value Chains
IFI	International Financial Institution
ILO	International Labour Organization
IMF	International Monetary Fund
ISP	Industrial Strategy Project
MC	Material Culture
MPE	Marxist political economy
NHS	National Health Service
OECD	Organisation for Economic Cooperation and Development
PPPs	Public private partnerships
PSSoP	Public Sector System(s) of Provision
SoP	System(s) of Provision
UNRISD	United Nations Research Institute for Social Development
WHO	World Health Organization
WRA	Welfare Regimes Approach

CHAPTER 1

Introduction and Background to the SoP Approach

1.1 Introduction

Across the social sciences, academics have long grappled with the question of what drives consumption. Economics has jostled with anthropology, sociology, psychology and other disciplines to account for who gets what, how and why and with what significance. These questions are important. Consumption relates not just to more or less optimal decisions about shopping habits but also to the way in which people's basic needs are met. Consumption of (and, indeed, access to) essential services such as healthcare, education, housing, transport and water has profound effects on human welfare. Food consumption affects public health but also intersects with mental health and is at the heart of many social interactions. The meanings and understandings attached to different forms of consumption are at the core of our engagement with the world. Consumption is therefore fundamental to our social as well as our economic and environmental development. What, then, determines what we consume? While neoclassical economics suggests that consumption decisions are individual, there is a wealth of theory and evidence to indicate that what is consumed, as well as how and why it is consumed, is associated with an extensive range of influences that shape what might be termed consumption norms, themselves embedded in societal dynamics, rooted in historical and geographical context.

The systems of provision (SoP) approach was developed thirty or more years ago in response to what were perceived to be the limitations of

K. Bayliss, B. Fine, *A Guide to the Systems of Provision Approach*,
https://doi.org/10.1007/978-3-030-54143-9_1

consumption studies across the social sciences whether taken individually or in conjunction with one another. The starting point for the approach is that the drivers of consumption are diverse, and outcomes derive from the details of the system within which specific commodities in particular, are both produced and consumed—whilst acknowledging that not all consumption derives from market access alone. There are distinctive systems, for example, for fashion and food, alongside those for housing and energy, and so on, each subject to its own specific mode of provisioning, distinct from others. It is the elements of these systems, and their intersections, which shape consumption outcomes. The SoP approach, therefore, sees consumption of particular commodities as a component of an integrated 'vertical' system, relating to the particular item of consumption under consideration, linked to processes of production, distribution and exchange (and sometimes disposal) as well as to 'horizontal' socio-economic variables such as income and gender. Also significant for consumption are the material cultures, or meanings, associated with commodities and these are also specific not just to the commodity but in time and place. They will be affected by the mode of provisioning, as for example, when it comes to designer versus popular high street fashion, even if the item is otherwise the same. The commodity in question is itself also significant, not subject to sweeping generalizations, theoretical or otherwise. Cultures, and so choices and meanings, associated with food, for example, will be different from those for water or housing. Context is also important. Water, for example, has different meanings for consumers in a flood as compared with a drought.

A penetrating understanding of who gets what and why would need to incorporate all of these elements. But complex conceptualizations are involved. The SoP approach has been refined over three decades drawing on academic disciplines across the social sciences to devise a framework for analyzing the elements that shape consumption outcomes. The theoretical framing and the empirical applications are the subject of this book. This chapter sets out the trajectory of the approach, from its initial conception, in accounting for the relatively narrow focus upon the widening use of consumer durables, to more recent applications in social policy and the grander issues of material cultures of financialization. The chapter shows that over time, the approach has proven to be remarkably robust. Moreover, there is potential for further contributions as the challenges we face appear increasingly to be systemic, from the multidimensional issues posed by climate change through to the more general prospects of what has come to be known as social reproduction, as shown in the chapters that follow.

1.2 The Changing State of Consumption Studies

Interest in the study of consumption has been prospering over the past thirty years across the social sciences. For example, postmodernist deconstruction, especially with its application to consumption within cultural studies and with Baudrillard (1981, 1988) to the fore, was positively thriving in interpreting the meanings of the consumed to consumers and the corresponding constructions of their individual identities. Such approaches were complemented by more longstanding, more broadly cast, and possibly more inflexible, approaches from within other disciplines—whether it be, for example, the utility maximization of homo economicus of mainstream economics, the emulation and distinction of sociology, the ethnographies of anthropology, the potential affects derived from psychology. The 'social' nature of consumption came into the study of marketing, management, business and advertising (Robertson 2020). For, alongside the more 'academic' perspectives sat what might previously be seen as the more or less mundane studies driven by the wide-ranging imperatives to produce, distribute and sell to consumers at a profit, in part by the moulding and manipulation of desires.

Subsequently, at one extreme of consumption studies stood the excitements and novelties of postmodernist deconstruction and, at the other, the dull insights from traditional social sciences. Inevitably, a middle road emerged, giving rise to attention to the ways we live our lives in general and how we consume in our everyday lives in particular. This brought, for example, globalization (McDonaldisation[1] even (Ritzer 1993)) of both consumption and its cultures to a position of prominence as an important theme, soon to be followed by neoliberalism and, most recently, financialization. Iconically prominent in both scholarship and everyday life came the temporarily ubiquitous Walkman (Du Gay, Hall, Janes, Mackay & Negus 1997). It served as the erstwhile symbol of the evolving state of consumption and consumer studies. Today, no doubt its counterpart is Apple for which its dependence on cheap, low wage electronic component manufacture (Albergotti 2019) is considerably less observed than the symbolic bitten apple itself. Indeed, as consumption itself was, as always, changing, so consumption studies began to move rapidly and fully away from postmodernist, discursive deconstruction to the reconstruction in light of the whos, the hows, the whys and the where froms of the provisioning of consumption, not just the meanings of the consumed to the consumer.

In the intervening years since consumer studies took off, then, there has been something of a reaction against the postmodernist approach, not so much to set aside an understanding of the cultures of consumption but more to root them in, and through interaction with, the material practices to which they are attached. Such changes have given rise to a field of study that can be termed the material culture of consumption. It is concerned with the ways in which consumption is shaped by the associations and meanings attached to specific commodities and with the origins and transformations of such cultures, discussed in detail in Chap. 3. An example of the changing nature of material cultures is the gentrification of beer in the UK and elsewhere, which began as a campaign for 'real' as opposed to mass-produced beer. Whilst it has embodied the proliferation of local breweries, it has increasingly steered towards the embourgeoisement of drinking practices and cultures, not unlike those associated with fine wines just as wine as mass-produced plonk has become universal, (Thurnell-Read 2018).

Of course, the gentrification of beer (and wine in its own way as it always was, as opposed to beer, gentrified previously in the UK), has its counterparts in other such gentrifications, not least housing. However, we begin to see that gentrification, in consumption or otherwise, tends to have its counterpart in the non-gentrified, as different types of housing tenure, for example, have developed their own material cultures from (the generally preferred) owner occupation through to social housing, discussed in more detail in Chap. 4. Across these examples and more, there has increasingly been a reaction against, or alongside, the cultural turn in the study of consumption. It reflects at least in major part a response to much wider contemporary developments and structural shifts themselves, with the notions and practices of globalization and neoliberalism increasingly being seen as coming to fore as major factors impinging upon our everyday, including our consuming, lives. Consumption studies in general turned towards the material from the postmodern, both reflected in and, in its own way, through the SoP approach. The shift to the material culture of consumption was, then, in major part prompted by both the acknowledgement of globalized neoliberalism and a reaction against the failures of postmodernism in this regard (although other post-postmodernisms were waiting in the wings ready to fly). The major exception has been mainstream, and much heterodox, economics in which the renewal of the study of consumption moved at most marginally away from exclusive reliance on the rational decision-making of individual utility maximizers by incorporating varieties of other behavioural assumptions on top (Fine 1996, 1997a and, for a recent retrospect, 2019).

Thus, in short, that the world might be changing significantly away from its post-war boom comfort zone was not something that could be understood by confining analysis to what was going on in our heads. Major structural shifts were taking place with neoliberalism and globalization, also feeding into what was going to be recognized as the financialization of economic and social reproduction, all of which impact on consumption in diverse and intersecting ways. Once we begin to unpick neoliberalism itself, alongside globalization, we find that it affects consumption through any number of channels—insecure incomes alongside widening inequalities, austerity, and privatization—quite apart from the shock delivered to consumption, on credit or otherwise, and ultimately by the global financial crisis and its aftermath.[2] These global structural shifts are in major part, seen to affect consumption, for example, through mortgaging of housing as well as through the practices and cultures of credit card consumption and indebtedness. And, on top of these challenges to the study of consumption has been the new preoccupation with the relationship between consumption and climate change and sustainability, although the ethics of consumption has been an enduring if shifting theme especially in the context of consumer politics (see Chap. 6).[3]

The study of consumption has evolved to some degree in response to these shifting intellectual fashions and contexts. Yet, although preoccupation with the material culture of consumption serves as a useful descriptive overview of the current state of the field, the fragmented and disparate character of the collective study of consumption is striking, with no predominant approach, discipline or focus currently to the fore. The result has been the (slightly awkward) coexistence of an array of theories and variables deriving from different disciplines. Each of these tends to have a different definition of what is consumption (and the consumed and the consumer) along with how it is to be understood and explained. For economists, for example, it is a question of quantity, price and expenditure; for the anthropologist, the ethnographic meaning displayed by the consumer and the consumed; for sociologists, consumption is an important part of the formation of individual and group identities, and so on for other disciplines. Like so many other topics, the study of consumption remains siloed, soiled even, within particular disciplines with at most token acknowledgement to interdisciplinarity. There are, interestingly, exceptions (and not the rule) but these tend to emerge in the context of case studies of particular items of consumption (such as car cultures) and/or particular generic aspects of consumption (such as male or female cultures of consumption, often treated separately from one another as opposed to mutually gendered).[4]

It was this collective failing of the social sciences to provide a coherent and robust approach to consumption that led to the development of the SoP approach in the first instance. It has remained a vital motivation ever since.

1.3 Evolution of the SoP Approach

The refined SoP approach that has emerged today is the result of more than three decades of work, inspired by dialogue with relevant literatures, topics and case studies, and punctuated by a series of more or less fortuitous research projects and programmes. This section sets out the ways in which the approach progressed by bringing together, and shedding light on, what make up its core components and how they have evolved and mutually interacted with one another. The details of these aspects are set out in more detail in Chaps. 4 and 5 which deal with SoP applications in practice.

1.3.1 Phase 1: Consumption Norms

The SoP approach dates back to a critical review of consumption theory which arose out of research in the early 1990s into the relationship between rising female labour market participation over the post-war period and its relationship to the rising ownership of consumer durables. The presumption within mainstream economics was that some sort of comparative advantage was at work, with (cheaper and more efficient) consumer durables releasing women from domestic labour from which they could more effectively be engaged in waged work instead. Wages from labour market participation were considered to more than compensate for the cost of labour-saving devices within the home (Fine 1992, 1995). However, such an approach was found to be wanting. Closer analysis indicated that the ownership of 'white goods' was increasing across the spectrum and was not necessarily associated with female labour participation, let alone the latter playing a causal role.[5] How could this be explained?

Consumer theory within mainstream economics since the 1870s had been based on the idea of fixed preferences for fixed individuals with a fixed and sole motive—utility maximization—over fixed goods effectively defined by physical properties, subjectively enjoyed in the most facile fashion without reflection upon consumption itself by the consumer. And yet, from the 1950s, despite its legion deficiencies in the realm of consumption itself, such consumer theory was in the process of being extended across ever more areas of economic and social life as part of the early phases of what has come to be known as 'economics imperialism' so that, for example, choosing between war and peace was seen as akin to choosing between apples and pears (Fine 2019). Close assessment of shifting patterns of ownership of consumer durables revealed that neoclassical economics was found wanting in terms of its own favoured terrain of supply and demand as well as in addressing *why* consumers have the preferences they do, what leads from those preferences to purchase and use, and what meanings and cultures are attached to corresponding activities of purchase and use.

The SoP approach in part arose out of looking for alternative accounts of consumption than the narrowly reductionist and individualistic theory derived from mainstream economics, and its crude attachment of (gendered) labour market patterns to (gendered) consumption patterns. Of course, consumer theory is not, and has not been, limited to the confines of mainstream economics. Significant for developing the SoP approach was to situate it in relation to wide-ranging contributions from across the social sciences. By contrast to the atomistic individual of economics, post-modernism had (re)constructed the consumer as subjectively flexible, inventive and more or less unbound by the material properties of the consumer and consumed themselves (Fine and Leopold 1993; and Fine 2013 for retrospect).

Exploring understandings of consumption further in this way, it emerged that there was a more or less unlimited array of theories and variables upon which to draw, especially once incorporating psychology, market and business studies and the burgeoning field of consumer studies itself, quite apart from the more established disciplines of sociology and anthropology. What these revealed from the perspective of the SoP approach was the presence of a range of 'horizontal' theories, typically bound within particular disciplines and marked by their methods and subject matter, ranging, for example, over emulation and distinction through to ethnographies. Could all such horizontal theories be stacked to give some sort of comprehensive synthetic approach to consumption? Fine and

Leopold (1993) took the position this would not be possible other than in indiscriminatingly accumulating a s(t)ack of mutually determining variables. These would need to be structured in relation to one another, conceptually and causally as well as in terms of degree of importance irrespective of whether corresponding horizontal factors could be rendered mutually compatible in light of their different disciplinary origins (and hence theories, methods, methodologies and conceptualizations). Your utility is my ethnography, is their emulation and distinction! This critical perspective on horizontal theories drawn from within and then across the disciplines fed into the SoP approach through specific rejection of the 'pick and mix' approach to the combinations of 'horizontal' approaches that characterized most of the interdisciplinary literature. This inconsistent bundling of variables was strongest in consumer or marketing studies where there was less of an interest in analytical coherence and more of an inclination to seek out an empirical model that might explain why consumers buy or how to induce them to do so.

In contrast, but drawing more constructively upon the literature, the SoP approach introduced a 'vertical' approach to the analysis of consumption, focusing on the chain of activities that connects production to consumption (and even disposal) across the life of the commodity. The idea was put forward that who produces, distributes, and sells, and what, how and to whom is differentiated from one commodity to another, with each commodity characterized by its own integral system of structures, relations, processes and agencies. The SoP approach deployed intra-disciplinary contributions by melding them together, if differentially, in the context of particular commodities. Depending on the research question, and the commodity in question, there would be a need for understandings, drawn across economics, anthropology and other social sciences as well as the physical sciences in some cases. Moving out of academic silos is at the core of the ethos of the SoP approach. Chapter 2 explores the theoretical underpinnings of the approach in more detail.

To return to the subject of consumer durables, the newly devised SoP approach indicated that, rather than individual decision-making, much more important for the study of consumption was to establish the presence of consumer *norms* around particular durables. These are to be understood as a greater or lesser disposition to own consumer durables according to socio-economic characteristics such as household composition, employment status, income and so on, without otherwise necessarily casting aside narrowly conceived economic factors such as price and income. Consumer durables in advanced countries, such as TVs, washing machines and central heating, have become standard irrespective of female labour market participation, although how such common ownership has come about reflects shifting norms across the population whose content and dynamic have themselves to be explained (other than as trickle-down from elites, for example). Other durables have different sorts of norms, with ownership contingent on household characteristics—as with (dual) car ownership for example. Consumption patterns are heavily gendered as well as being differentiated by socio-economic, ethnic and other variables attached to status. In addition, and significantly for subsequent work, the research on consumer durables allowed for a considerable review of labour market theory, the family and female labour market participation,[6] and, inevitably, consumer theory as a whole.

This is not to deny that expanding female labour participation (especially of married women with children) coincided with the expansion of consumer durables, or ready-made foods for example, as putatively labour-saving devices. These have unambiguously influenced both patterns and content of consumption as well as shifts in the what and how of household roles and activities (from microwaves to entertainment devices). However, such effects cannot be reduced to simplistic accounts of the intersections of utility maximization in response to changing production possibilities and technologies, not least because the nature and identities of consumers (even if represented by utility functions) are themselves subject to transformation, as well as reflecting inequalities and differences by gender and race, for example. The relationships underpinning norms of consumption by social status are far more nuanced and contextual.

As a result, when consumption is viewed through the lens of consumer norms and who gets what and how, questions arise over what form such norms might take, how they become established and how they might be changed. The SoP approach, then, was initially steered towards theoretical and empirical understanding of the systemic, if commodity-specific, origins of consumer norms—how both to identify and to explain them. Especially influential in this respect, over and above work on consumer durables and female labour market participation, was the work on the UK housing system by Michael Ball (1983, 1988). Ball argued that it was insufficient to focus on housing finance (the terms and conditions of access to a mortgage) to understand housing access or consumption even if complemented by analysis of different forms of housing tenure (social or private rented, or owner-occupied). Rather these elements needed to be set within the context of how housing supply is subject to planning, access to land and building supplies, and the organization of the building industry itself.

Such insights are not confined to the housing system alone but apply equally to the clothing, food, energy, transport and other systems which furnish consumption. Each can only be usefully addressed through identification of differentiated SoPs organized along specific commodity lines. Such specifics involve tracing back from consumption and consumers through to the material practices by which provisioning is systemically reproduced and transformed as a whole. These insights informed the classic presentation of the SoP approach in Fine and Leopold (1993), that also offered case studies (especially of clothing and food), alongside historical controversy (over the putative consumer revolution of the eighteenth century or, indeed, the so-called 'consumerism' as definitional of any period).[7] Fine and Leopold also sought to reconstruct the contribution of other, intra-disciplinary, generally horizontal, theories of consumption from within the SoP perspective by seeing them as differentially relevant and to be incorporated into specific items of consumption, as appropriate. For example, is consumption gendered and if so how and why? When is there trickle-down or even trickle-up? Why does status play a role in some consumption but not in all, and when and why does this change? And so on.

These concerns are especially relevant in case of advertising as a horizontal factor, for example, for which the SoP approach offered a particularly acute contrast with, and criticism of, postmodernist accounts. For the latter, the activities of chocolate confectionary manufacturers were particularly of interest. In the 1980s and early 1990s, advertising for Cadbury's Flake took the form of exotic depictions of a woman eating a chocolate bar in various ways, including lying luxuriously within a Victorian standalone bath. For the postmodernists, consumption, or its representation or meaning, was deemed not necessarily to be connected to production at all and only weakly to the product itself in some cases. However, at the same time as Cadbury's were promoting its Flake, the most important source and growth of advertising expenditure in the UK came from supermarkets engaging in 'store wars' with promotions along the lines of 'come to us for your weekly shop because we are cheaper', with emphasis on own-label as opposed to branded goods. The relative attraction of this realist type of advertising was related to increasing intensity of competition between the highly concentrated supermarket sector, in part resulting from property developments in out-of-town shopping centres. Such competition was itself further intensified by a collapse in the then property boom which had rendered the major retailers on the verge of bankruptcy in light of expenditure on out-of-town sites in anticipation of building more hypermarkets.

In this context, supermarkets had to balance the attraction of cheaper own brands against consumer satisfaction from, and demand for, branded goods. (To be attracted) as a shopper, I might have to have a Flake. But as Flake manufacturer, I have to get my product on the shelf in place of, or alongside own label confectionary (although the two products might even be totally or substantially the same, other than the wrapper). To get product placement within supermarkets, branded goods had to be of sufficiently high demand to justify being sufficiently desired as part of the weekly shop at the expense of, or alongside, supermarket 'own brand' goods, thereby explaining the combination of the exotic, branded, more expensive, advertised goods alongside less imaginative own-label advertising appealing to cheaper price. For the SoP approach to advertising, the consumption story is about far more than fantasy advertising. Instead, it required attention to the more mundane matters of property markets, the weekly shop and the strategic targeting by both supermarkets and manufacturers of cheap and own label versus expensive and branded. How and whether this was done differed across products, whether milk, meat, bread, chocolate, cereals or toothpaste or other non-food household items (Fine 1996).

Such considerations extend to other horizontal factors, with differentiation in the extent and nature of the gendering of consumption, just as there is of its advertising—note how the (exclusive) masculinity-linked advertising of the one-car household has changed as more women have become drivers (and owners) in their own right. The SoP approach, then, in part responded to a gap in consumption studies connecting the cultures created by fantasy advertising to the material processes of how goods are produced and consumed.

1.3.2 *Phase 2: Food Systems*

The SoP approach took a leap forward in the late 1990s in light of an ESRC-funded project researching the UK Nation's Diet, seeking to understand why consumers were not following healthy eating guidelines. Reviewing assessments of food consumption, Fine, Heasman & Wright (1996) found a burgeoning discipline of food studies, unsurprisingly ranging across as many topics, variables and disciplines as consumption studies more generally. Whilst food system theory was alive and well (if beginning to go through a temporary crisis of confidence, see Chap. 5), a major task comprised how to bring together the broadening analytical range around the study of what we eat, and why while, at the same time, forging a framework for studying the food system (or systems) as distinct from other systems furnishing consumption whether housing, clothing or whatever. There was no differentiation for studying food consumption as an integral whole. It was argued by Fine, Heasman & Wright (1996) that the SoP for food needs to be addressed as a totality from production through to distribution and retailing, as well as corresponding consumption.

Applying the SoP framework empirically in the same way as for consumer durables, attention was focused on the meat, sugar and dairy *systems*—although now the motivation was to explain the nation's (poor) diet, rather than female labour market participation. For dairy, in particular, the rise of supermarket retailing had ensured the availability of a wider range of healthier low fat milks, once the uniquely protected doorstep delivery was effectively abandoned. But with the agricultural systems supporting production of high fat milk, the skimmed off cream had to go somewhere. And so it did, into fancy cheeses and desserts and manufactured foods all equally readily available in the multi-product supermarket (with, ironically, those taking low fat milk from one counter, or refrigerated cabinet, also tending to take the fancy cream products from another). A similar story can be told for sugar with reduction in direct consumption from the sugar bowl being compensated for by its incorporation within manufactured foods (alongside salt and unhealthy fats), sustaining its level of consumption per capita. Thus, analysis of the dairy and sugar systems indicated that healthy eating programmes for the consumer would tend at most to redistribute consumption and most likely towards those on low incomes and poor diets in the first place, and least able or willing to respond to health messages (Fine, Heasman & Wright 1996, 1998 and see Fine 1998a for more on sources for these conclusions).

This analysis led Fine to explore further the cultural understandings of consumption and the ways in which these intersect with provisioning systems. His findings demonstrate that there are limits within which consumption and its meanings to consumers in terms of health beliefs (and actions upon them) are bound by the functioning of the respective food systems. There are also implications for the definition of the boundaries of the food systems themselves, with dairy taken as a whole rather than just milk, sugar including its presence in processed food although not in artificial sweeteners. It became clear that the analytical boundaries in determining how consumers choose should not be limited to the point of consumption itself but need to be traced outwards to the backward linkages to, and determinants of, consumption (Fine, Heasman & Wright 1996; Fine 1998a).

These findings had important implications for understanding the nation's diet, and the norms of consumption attached to it, especially in the context of the rapidly growing incidence of dietary diseases associated with overeating. Campaigns to promote healthier consumption had focused on the choices made by the consumer in closing the gap between the actual and ideal diet. But these had failed to address the systemic issues of *what* gets produced and, hence, distributed and consumed. Also significant is the *way* in which food is consumed, for example, with a shift towards out-of-town supermarkets, increasing ownership of microwaves and fridge/freezers, and the proliferation of cafes and snack foods as well as ready-made meals. The SoP approach rejected the idea that the targeted ideal diet would motivate consumers, especially using this strategy alone, because the determinants of consumption are as varied as the SoPs to which they are attached. Dietary ideals play at most one part amongst many in determining what is consumed and the role of consumer choice/culture within it. This conclusion has had widespread implications for understandings of eating disorders, discussed further in Chap. 5.

1.3.3 Phase 3: PSSoP

Until the 2000s, the SoP approach was largely confined to studies of private, commercialized consumption, not least in taking existing studies as its critical point of departure. There was, however, a question as to how it might be applied to the realm of public sector provisioning, or consumption outside of the market. Applying the SoP approach to modes of public provision gave rise to what Fine (2002) termed the public sector SoP, or (the somewhat clumsy acronym) PSSoP, where the object of study is still consumption but production is typically in the public sector and/or where there is state intervention to promote social outcomes. This expansion of the SoP approach was motivated by a number of factors: first, to acknowledge the significant role of the state even where provisioning is primarily private (e.g., as with housing); second, to rectify the extent to which public as opposed to private consumption had been neglected in the literature (see Chaps. 4 and 6 for the reasons for this); and, third, to study and offer alternatives to the drive to privatize, commercialize, etc. This was first explicitly acknowledged in the new edition of *World of Consumption*, Fine (2002) but see also Fine (2005a). It is not just the terminology that is clumsy but the more or less arbitrary division between what is a SoP and what is a PSSoP, especially as privatization has taken many different forms and there is such variation in levels of private and public sector participation. Nonetheless, the distinction has enabled the SoP approach to engage much more broadly, not least in embracing social policy. Consumption of publicly provided goods and services has often taken the distinctive form of welfare provision and often been perceived other than as consumption. This was at least until the languages and practices of the commercial sector became introduced to a greater or lesser extent into the public sector as part and parcel of neoliberalization and commodification, blurring the boundaries between public and private.

As with consumption, a theory of social policy must accommodate a variety of determinants, how they interact across structures, agencies, processes, relations and corresponding institutions to give rise to a diversity of shifting outcomes. With increasing privatization of social provision, the SoP approach provided a systematic way of understanding consumption and production of welfare services across different types of supply. End users (effectively being reconstructed as 'consumers') engage with public and private providers. The conceptual gaps in mainstream consumption theory were found to apply equally, if not more so, to provision within the public sector (Fine 2005a, b but also closing chapter of Fine 2002).

The SoP approach provided a useful entry point for the study of the failings of the hugely prominent welfare regimes approaches (WRA) deriving from the work of Esping-Andersen (Esping-Andersen 1990, 1999, 2009). Fine (2014) develops a critical, SoP-inspired, approach to understanding the problems of the WRA, in response to, and in explanation of, the apparent neglect in mainstream consumption studies of public consumption.[8] The WRA, in particular, is found to be insufficiently sensitive to differences across different sectors of social provisioning (and negligent of the role of financialization and its underpinnings of neoliberalism over the past two decades or more, in common with other mainstream approaches to social policy). Efforts to organize the diversity of the national approaches to welfare policy, and its different components, adopted by different (nation) states resulted in a burgeoning number of welfare regimes and typologies with limited classificatory coherence (Fine and Bayliss 2016). Further, as soon as consumption becomes recognizably public (and struggled for as such) it has tended to be redefined as something else, most notably the welfare state or social policy (see Chap. 4). This places it outside the realm of consumption studies, yet the issues of asking who gets what and how and why are the same. In mainstream literature on welfare regimes, effectively all government provision tended to be seen within a dualism of being equivalent, to a greater or lesser extent, to private provision (is social housing as good as private renting or owner-occupation) or seen as distinct from (private) consumption altogether by being alternatively designated as social policy and/or as belonging to the welfare state. However, the PSSoP framing highlighted that, as with the SoP approach in the context of commodity consumption, understanding of social policy is highly context- and sector-specific, requiring an integral, holistic framing rather than either a focus on a particular aspect such as the presence or absence of user charges or general, universalizing theorizing.

The themes attached to PSSoPs were developed further as part of a five-year EU-funded research programme, Financialisation, Economy, Society and Sustainable Development (FESSUD),[9] that began in 2011. The aim was to explore the application of the evolving SoP approach to the 'consumption' of basic (public) services, situating these within the wider context of financialization. Parallel case studies were conducted to analyze the SoPs for water and housing across five countries: the UK (Bayliss 2014; Robertson 2014), South Africa (Bayliss 2016a; Isaacs 2015), Portgual (Teles 2015a, b; Santos, Serra & Teles 2015), Poland (Lis 2015a, b) and Turkey (Çelik, Topal & Yalman 2015; Yilmaz and Çelik 2015), with synthesis analyses of the cross country findings (Bayliss 2016b; Robertson 2016). Additional empirical research took SoP into pensions (Churchill 2013) and health (Bayliss 2016c). The programme also allowed for the refinement of the theoretical and methodological principles for SoP, discussed further in Chaps. 2 and 3, and see (Bayliss, Fine & Robertson 2013).

In terms of the evolution of the SoP approach some core themes emerged from the FESSUD studies. First, common across sectors and locations was the penetration of neoliberalism and financialization, but the extent and effects of this differed widely, depending on the interrelations between the core SoP elements. Diversity emerged not just across location but over time across institutional structures and policies.

Second, SoP has come full circle with regard to the (material) cultures of consumption. As already indicated, the SoP approach was originally motivated in part by antipathy to the postmodernist inclination to deconstruct cultures of consumption without regard to the material processes involved in provisioning (and, possibly, even material consumption itself given focus upon meanings as opposed to practices of consumption). Having (re)introduced provisioning to the study of consumption, attention turned within the SoP approach to (re-)engage more constructively with the material cultures themselves, see Chap. 3 (although this had always been present as illustrated above by studying the diet paradox, eating disorders and the nature of advertising, see Chap. 5).

Third, the FESSUD research shed light on distributional outcomes, emerging from the systems by which essential needs were met, and the narratives and discourses by which they were sustained. For example, the reconfiguration of housing from a form of shelter to a financial asset led to specific patterns of consumption which penalized the most marginalized, although this varied widely across the case study locations. In the provision of water, privatization in England has created revenue streams which have in some cases rewarded wealthy shareholders, financed by household bills. Understandings of social policy were similarly varied across location and over time. However, there were common themes. For example, previous aspirations of universalism have been increasingly displaced by residualism with state provision as a last resort (in housing) and greater discrimination between 'deserving' and 'undeserving' poor, as well as state support increasingly directed at sustaining financialized private provision. See Chap. 4 for more on these case studies.

Overall, the FESSUD body of research drew attention to the systemic, yet heavily contested, contextualized processes by which cultures, institutions, policies and practices have supported specific constellations of powerful interests, from the so-called minerals-energy complex in South Africa to global finance in the UK. In many respects, the findings, for example, regarding the neoliberalizing of economic and social reproduction were far from unique. However, locating these in their case-specific settings highlighted the complexity and diversity of households' engagement in these processes in ways of which they are mostly unaware (see Chap. 4).

The research programme led to new directions in the study of financialization including a special volume on material cultures of financialization. It is widely established that finance is drilling down to, and profiting from, many aspects of everyday life in contemporary capitalism. The contribution of the Special Issue was to examine how such extraction has become normalized, through shifting narratives, shaping material cultures (Fine 2017a; Bayliss, Fine & Robertson 2017, and see Chap. 4). Fine (2017a) draws an analogy between financialization and food in that households are under the contradictory pressures both to save (as to diet) and to spend (on credit) (as to eat). Each of these imperatives to invest and to save is heavily promoted by financial services and mutually reinforce each other so households feel the pressures both to save and to spend simultaneously. The expanded presence of finance in social reproduction has transformed agents' subjectivities. The ways in which this encroachment has infiltrated all aspects of life have been facilitated by the ideas and meanings attached to finance, as captured in the material cultures of financialization.

1.4 The SoP Approach's Durability

The SoP approach has gone from strength to strength since its inception in the early 1990s. It emphasizes, as already indicated, the vertical integration of production with consumption, alongside the (society-wide) horizontal factors determining and contextualizing consumption. In this respect, it has come to accept some traditional methods and concepts from across the social sciences, exploring specifically how relations, structures, agencies and processes of provisioning are (re)constituted alongside the (material) cultures with which they interact. As shown above, the approach was developed to respond to material developments in consumption itself and their interaction with wider social and historical phenomenon. As a result, it was, as it were, ahead of the game in terms of the material culture of consumption that was about to react against the extremes of the post-modernist, interpretive turn. Moreover, SoP has adapted its own distinctive way of addressing material culture itself, with applicability beyond consumption alone. Indeed, the privatization of public services (health, education, utilities, etc.) has posed a challenge in extending the scope of consumption studies, as such services are typically seen as welfare provision. The SoP approach, however, has been able to adapt to develop the notion of (public) *consumption* to provide a systematic form of study of different types of provisioning—both within and across public and private sectors.

The SoP approach has the capacity to address and incorporate what might be termed epochal changes and universal and global determinants, such as commodification of production, globalization, neoliberalism and financialization. In addition, the approach addressed the more specific and contextual determinants attached to particular consumers or objects and practices of consumption (from the global to the household, or even the pub to revisit real ale). Importantly, the approach connects the high level with the local, and the macro with the micro. Given the stance that each SoP is itself distinctive, it follows that this cannot be done through a general theory of consumption although general theory as such, as opposed to its contextual application, is not eschewed. The biggest contrast in this framing is found with the mainstream economics' acontextual notion of consumption as demand determined by individual utility maximization in light of income and price constraints.

The SoP approach has been designed to forge, even to dissolve, interdisciplinarity. Although grounded in political economy as a starting point, the approach draws critically if constructively upon the existing literatures but, in large part, in reaction against the 'horizontal' framing of consumption studies within disciplines. Hence, just as there is no general theory of consumption, so there can be no general theory of SoPs in the sense that the application of the approach will be dependent upon the particular system under consideration. The theoretical and analytical issues raised by a health system are liable to be very different and differently addressed than for the food system, with different focuses for attention as well. Appropriate theory will be SoP-specific, depending on what is studied, how and for what reason. For example, research into environmental outcomes will require a different focus from a study of the impacts from financialization or upon inequality. Equally, the SoP approach offers the opportunity for comparative case study, across different SoPs, or for the same SoPs across different times or locations, in which more than simple empirical comparison is involved. Understanding the nature of, and reasons for, differences between SoPs requires a combination of theoretical and empirical analysis through application of case studies. Chapter 2 sets out theoretical concepts on which the SoP approach was built, which draws heavily on Marxist political economy (MPE) and in turn links to the creation of both use and exchange value. The chapter distils, as indicated, an extensive body of work in this field down to five fundamentals that should be part of SoP research: agents, relations, processes, structures and material cultures.

Within this framework, there is a degree of openness in the methodology, methods, theories and concepts that are deployed. By analogy, the Entitlement Approach to famine made famous by Amartya Sen (1981) draws attention to who gets access to adequate food in varying circumstances in opposition to the Food Availability Decline hypothesis. In other words, it points to how famine may result without a fall in food supply but without necessarily specifying how and why this occurs, either theoretically or empirically—although much and varied work of this sort has been inspired by the Entitlement Approach, much of it emanating from Sen himself. Nonetheless, whilst this and the SoP approach are open, they are not a blank page, not always starting from scratch in which anything goes. The SoP approach has its own precedents in terms of both theoretical determinants and case studies, that set traditions, just as is the case with and from Sen's work. Indeed, from the perspective of the SoP approach, the entitlement approach has been critically assessed both for failing to be specific about the food system as such (entitlements could be about things other than food) and for grounding a social outcome (famine) in individual entitlements (famine is not just a lot of individuals failing to get insufficient to eat) (Fine 1997b).

As a result of the above features, the SoP approach provides a robust basis for both making policy and analyzing its effects as it is grounded contextually and systemically in its analyses of the real worlds of consumption. This is in contrast with, for example, the abstract notions of an assumed environment of more or less perfect markets and optimizing individuals which dominates mainstream economics. Being rooted in such realities is a significant strength and offers considerable insights. For example, rather than assuming that interactions between agents are benign, the approach takes the view that agents are likely to have conflicting objectives and that allocations will be contested. Such contestation will emerge in ways that reflect, and give rise to, different understandings of consumption itself. Furthermore, clearly not all agents have equal power to shape the SoP, so unpicking the details of such relations is fundamental to understanding how outcomes (can and will) emerge.

In addition, when it comes to social policy, mainstream interventions are typically focused on the needs of the household with corresponding measures to support consumption of those poorly served, for example, confining attention to (poor) individual decision-making as well as degrees of access and affordability. In contrast, SoP analysis also requires attention to, indeed grounds such issues in, the whole system linking consumption with production. Such insights reveal the role of the production process (and other factors distant from consumption itself) in creating specific social outcomes. Furthermore, with attention to the material cultures which underpin consumption, the approach is able to analyze the understandings that drive policy interventions and the impacts of these on the consumer in terms not only of what is consumed but how it is received and, possibly, contested. In short, according to Fine (2013, p. 34):

> the (PS)SoP approach has the advantage of potentially incorporating each and every relevant element in the processes of provisioning, investigating how they interact with one another, as well as situating them in relation to more general systemic functioning. This allows for an appropriate mix of the general and the specific and, policy-wise and strategically, signals where provision is obstructed, why and how it might be remedied.

The purpose of this book is both to set out the elements of the SoP approach and to provide examples of its application. Hence, the contents fall into two parts. The following two chapters focus on the conceptual and theoretical aspects of the approach. Chapter 2 details the underlying literature on which the approach is founded, notably drawing on Marxist political economy, and lays out the core components of SoP research. Chapter 3 is devoted to the theme of material culture and the ways in which meanings and understandings are shaped. Identifying material cultures, and the forces by which they emerge, is challenging as these are often deeply ingrained. To assist in this endeavour, the chapter provides a list of ten elements of material cultures, known as the 10Cs.

The second part of the book looks at applications of the SoP approach. These have been many and diverse and so only select themes can be covered in depth. Chapter 4 brings together some of the SoP contributions in the area of social policy and social reproduction, in particular around the everyday, looking at consumption of housing, water, health services and fashion. This grouping clearly demonstrates the material diversity in what is consumed, how it is provided, and the meanings with which these goods are associated, highlighting the importance of a context-specific approach to consumption.

Chapter 5 provides a comprehensive account of the SoP for food, linking the emerging epidemic in obesity with underlying provisioning systems. The final chapter sets the SoP approach in context, situating it within broader literature in consumption studies. As a corollary of the SoP approach's ability to span the most abstract and the more specific, it has also been enabled, as already indicated, to keep abreast with contemporary developments, not least with the neoliberalization (and financialization) of consumption to the fore, however these might be understood. Going forward, this is especially important in light of, for example, the increasing interest in the relationship between consumption and climate change and sustainability more generally.

1.5 Conclusion

SoP was described over two decades ago by Leslie and Reimer (1999, p. 405) as 'perhaps the most comprehensive elaboration of production consumption relations', and as has also been seen as one of the main approaches to the study of consumption, cited as such in Jackson, Russell & Ward (2004, p. 8). For Mansvelt (2005, p. 111) 'Systems of provision chains have provided valuable insight into commodity trajectories in spatial and social contexts. They emphasize the contexts in which commodities acquire form and meaning, as a way of demonstrating how things are embedded in social life and of exploring the social life (or commodity biographies) of things'. From an unexpectedly mainstream quarter, the OECD (2002, p. 8)

> The systems of provision approach analyses consumption as an active process, with actors seeking certain lifestyles, and constructing their identity by selective consumption and practices. The "systems of provision" is defined as the chain that unites particular systems of production with particular systems of consumption, focusing on the dynamics of the different actors (producers, distributors, retailers as well as consumers). In this light, it becomes clear that by the way governments design and transform energy, water and waste systems can either enable or obstruct household behaviour towards sustainable consumption.
>
> The systems of provision framework for understanding consumption patterns stresses the importance of exploring the mechanisms that shape everyday practices related to commodities and services and the extent to which they can be seen to support or impede sustainable consumption behaviour. In this light, household consumption is not the sum of individual behavioural patterns, each consciously motivated and evaluated by the actor. Instead, household consumption is a whole set of behavioural practices that are common to other households ... They are social practices carried out by applying sets of rules and shared norms. They are also connected to production and distribution systems (technological and infrastructure network) that enable certain lifestyles that connect consumers to one another.

The SoP approach has also formed part of a number of research programmes including i-Build (Infrastructure BUsiness models, valuation and innovation for Local Delivery) researching infrastructure interdependencies in the UK and developing new business models to improve delivery. According to one of the core iBuild project reports (Brown and Robertson 2014, p. 92):

> the SoP approach has the potential to deepen our understanding of the socioeconomic dimensions of infrastructure provision and should be given a prominent role in such future work. In its conceptualisation of agents and the economy, and its methodological commitment to starting with concrete conditions rather than abstract theory, the SoP approach differs radically from standard economics and this puts it in a unique position to shed light on certain key aspects of infrastructure provision.

The reach of SoP is broadening across sectors and across continents. There are numerous and growing applications of the SoP approach across diverse areas from meat consumption in Vietnam (Hansen 2018) to jeans and fast fashion (Brooks 2015) (see Chaps. 4, 5 and 6). More recently the Leverhulme-funded project, Living Well Within Limits (LiLi), at the University of Leeds, has adopted the SoP approach to connect environmental sustainability with household well-being through energy consumption. The project includes case studies of energy SoPs to link biophysical inputs and social outcomes. The rationale for deploying SoP as an element of this project is a desire to unpack the 'black box' of energy consumption in climate change debates. Thus, rather than simply focusing attention on drivers of consumption or sources of generation, this project is seeking to connect energy consumption with production processes. These are situated within broader structures that vary across location and over time. Two applications of the SoP approach within this project include the political economy of car dependency (Mattioli, Roberts, Steinberger & Brown 2020) and a contribution to the UN Special Rapporteur on Extreme Poverty's report on the UK, focused on the role of privatization of essential services of water, electricity and transport (Bayliss and Mattioli 2018).

This volume aims to consolidate diverse writings on SoPs bringing together a comprehensive account of the approach and its core elements in theory and in practice. We set out what has been done by SoP and how others might continue to develop and use the approach in future.

Notes

1. For a spirited attempt to keep the thesis alive as prosumption in the digital age, a generation later than it was first put forward, see Ritzer and Jurgenson (2010).
2. Not surprisingly, by sharp contrast with other disciplines and topics, the concept of financialization, although highly prominent in heterodoxy, has been notably absent from mainstream economics where it is most needed

but ill fits (Dymski (2015) and Mader, Mertens & van der Zwan (2020)). Inevitably, again in contrast to other disciplines, this means the relationship between financialization as such and consumption has also been absent.

3. Whatever the merits of doing so, and with deep-rooted traditions, much literature tends to revolve around consumer sovereignty or manipulation, false and real needs, and so on.
4. Significantly, the SoP study of consumption was gendered from the outset, given its concern for the impact of consumer durables on female labour market participation. See below.
5. For the empirical work on consumer durables, see Fine, Foster, Simister & Wright (1992a, b, c, d, e; 1993), Fine and Simister (1995) and Fine (1983).
6. See Fine (1992, 1995, 1998b). On durables themselves, see references in the previous footnote.
7. See also Fine and Leopold (1990).
8. See also Fine (2016, 2017b).
9. 'Financialization, Economy, Society and Sustainable Development' see fessud.eu.

References

Albergotti, R. (2019). Apple accused of worker violations in Chinese factories. *The Washington Post*. September 9. Retrieved from https://www.washingtonpost.com/technology/2019/09/09/apple-accused-worker-violations-chinese-factories-by-labor-rights-group/.

Ball, M. (1983). *Housing policy and economic power: The political economy of owner occupation*. London: Methuen.

Ball, M. (1988). *Rebuilding construction: Economic change and the British construction industry*. London: Routledge.

Baudrillard, J. (1981). *For a critique of the political economy of the sign*. St Louis: Telos Press.

Baudrillard, J. (1988). *Selected writings*. London: Polity.

Bayliss, K. (2014). *The financialisation of water in England and Wales*. FESSUD Working Paper Series, No. 52. Retrieved from http://fessud.eu/wp-content/uploads/2015/03/Case-study-the-financialisation-of-Water-in-England-and-Wales-.

Bayliss, K. (2016a). *Neoliberalised water in South Africa*. FESSUD Working Paper Series, No. 204. Retrieved from http://fessud.eu/wp-content/uploads/2015/03/FESSUD_WP204_Neoliberalised-Water-in-South-Africa.pdf.

Bayliss, K. (2016b). *The system of provision for water in selected case study countries*. FESSUD Working Paper Series, No. 194. Retrieved from http://fessud.eu/wp-content/uploads/2015/03/FESSUD.

Bayliss, K. (2016c). *The financialisation of health in England: Lessons from the water sector*. FESSUD Working Paper Series, No. 13. Retrieved from http://fessud.eu/wp-content/uploads/2015/03/Financialisation_Health_England_WorkingPaper131.pdf.

Bayliss, K., Fine, B., & Robertson M. (2013). *From financialisation to consumption: The systems of provision approach applied to housing and water*. FESSUD Working Paper Series, No. 2. Retrieved from http://fessud.eu/wp-content/uploads/2013/04/FESSUD-Working-Paper-021.pdf.

Bayliss, K., Fine, B., & Robertson, M. (2017). Introduction to special issue on the material cultures of financialisation. *New Political Economy, 22*(4), 355–370.

Bayliss, K., & Mattioli, G. (2018). Privatisation, inequality and poverty in the UK: Briefing prepared for UN Rapporteur on extreme poverty and human rights. Sustainability Research Institute Working Paper Series, No. 116. Retrieved from https://www.see.leeds.ac.uk/fileadmin/Documents/research/sri/workingpapers/SRIPs-116.pdf

Brooks, A. (2015). *Clothing poverty: The hidden world of fast fashion and second-hand clothes*. London: Zed Books.

Brown, A., & Robertson M. (2014). *Economic evaluation of systems of infrastructure provision: Concepts, approaches, methods*. iBuild Project Report. Retrieved from https://research.ncl.ac.uk/ibuild/outputs/reports/9940_iBuild_report_print_version%20WEB.pdf.

Çelik, Ö., Topal, A., & Yalman G. (2015). *Finance and system of provision of housing: The case of Istanbul, Turkey*. FESSUD Working Paper Series, No. 152. Retrieved from http://fessud.eu/wp-content/uploads/2015/03/Housing_Istanbul_WP152-FESSUD.pdf.

Churchill, J. (2013). *Towards a framework for understanding the recent evolution of pension systems in the European Union*. FESSUD Working Paper Series, No.12. Retrieved from http://fessud.eu/wp-content/uploads/2013/04/Towards-a-framework-for-understanding-the-recent-evolution-of-pension-systems-in-the-European-Union-FESSUD-working-paper-12.pdf.

Du Gay, P., Hall, S., Janes, L., Mackay, H., & Negus, K. (1997). *Doing cultural studies: The story of the Sony Walkman*. London: Sage.

Dymski, G. (2015, September 10). *How to engage with the global crisis: The shape of jazz to come*. Presentation at International Initiative for Promoting Political Economy Annual Conference, University of Leeds. Retrieved from http://iippe.org/wp/wp-content/uploads/2015/10/Dymski-IIPPE-keynote-as-presented.pdf.

Esping-Andersen, G. (1990). *The three worlds of welfare capitalism*. Princeton: Princeton University Press.

Esping-Andersen, G. (1999). *Social foundations of postindustrial economies*. Oxford: Oxford University Press.

Esping-Andersen, G. (2009). *The incomplete revolution*. Oxford: Polity Press.

Fine, B. (1983). The order of acquisition of consumer durables: A social choice theoretic approach. *Journal of Economic Behaviour and Organisation, 4*, 239–248.

Fine, B. (1992). *Women's employment and the capitalist family*. London: Routledge, 1992, pp. vii and 249; reprinted as Routledge Revival, hardback, 2011, and paperback, 2012.

Fine, B. (1995). Reconsidering 'household labor, wage labor, and the transformation of the family'. *Review of Radical Political Economics, 27*(1), 107–125.

Fine, B. (1996). From political economy to consumption. In D. Miller (Ed.), *Acknowledging consumption* (pp. 127–163). London: Routledge.

Fine, B. (1997a). Playing the consumption game. *Consumption, Markets, Culture*, 1(1), 7–29. Reproduced in A. Warde (Ed.). *Consumption*. London: Sage Publications.

Fine, B. (1997b). Entitlement failure? *Development and Change, 28*(4), 617–647.

Fine, B. (1998a) *The political economy of diet, health and food policy*. London: Routledge, reprinted as Routledge Revival, 2013.

Fine, B. (1998b). *Labour market theory: A constructive reassessment*. London: Routledge, pp. 306, reprinted in paperback, 2010.

Fine, B. (2002). *The world of consumption*. London: Routledge.

Fine, B. (2005a). Addressing the consumer. In F. Trentmann (Ed.), *The making of the consumer: Knowledge, power and identity in the modern world* (pp. 291–311). Oxford: Berg.

Fine, B. (2005b). Social policy and development: Social capital as point of departure. In T. Mkandawire (Ed.), *Social policy in a development context* (pp. 80–96). Basingstoke: UNRISD and Palgrave Macmillan.

Fine, B. (2013). Consumption matters. *Ephemera, 13*(2), 217–248.

Fine, B. (2014). *The continuing enigmas of social policy*. United Nations Research Institute for Social Development Working Paper Series, No. 2014-10. Retrieved from http://www.unrisd.org/unrisd/website/document.nsf/(httpPublications)/30B153EE73F52ABFC1257D0200420A61?OpenDocument.

Fine, B. (2016). The systemic failings in framing neo-liberal social policy. In T. Subaset (Ed.), *The great financial meltdown: Systemic, conjunctural or policy created?* (pp. 159–177). Cheltenham: Edward Elgar.

Fine, B. (2017a). The material and culture of financialization. *New Political Economy, 22*(4), 371–382.

Fine, B. (2017b). The continuing enigmas of social policy. In I. Ye (Ed.), *Towards universal health care in emerging economies: Opportunities and challenges* (pp. 29–60). London: Palgrave Macmillan.

Fine, B. (2019). Economics and interdisciplinarity: One step forward, N steps back? *Revista Crítica de Ciências Sociais, 119*, 131–148.

Fine, B., & Bayliss, K. (2016). *Paper on theoretical framework for assessing the impact of finance on public provision*. FESSUD Working Paper Series, No. 192.

Retrieved from http://fessud.eu/wp-content/uploads/2015/03/FESSUD_WP192_Theoretical-Framework-for-Assessing-the-Impact-of-Finance-on-Public-Provision.pdf

Fine, B., Foster, N., Simister, J., & Wright J. (1992a). *Consumption norms, diffusion and the video/microwave syndrome*. SOAS Discussion Paper, No. 19.

Fine, B., Foster, N., Simister, J., & Wright J. (1992b). *Access to phones and democracy in personal communication: Myth or reality?* SOAS Discussion Paper, No. 20.

Fine, B., Foster, N., Simister, J., & Wright J. (1992c). *Who owns and who wants to own a car? An empirical analysis*. SOAS Discussion Paper, No. 21.

Fine, B., Foster, N., Simister, J., & Wright J. (1992d). *Consumption norms: A definition and an empirical investigation of how they have changed, 1975–1990*. SOAS Discussion Paper, No. 22.

Fine, B., Foster, N., Simister, J., & Wright J. (1992e). *Consumption norms for durables: Evidence from the general household survey*. SOAS Discussion Paper, No. 23.

Fine, B., Foster, N., Simister, J., & Wright, J. (1993). Consumption norms, trickle-down and the video/microwave syndrome. *International Review of Applied Economics, 7*(2), 123–143.

Fine, B., Heasman, M., & Wright, J. (1996). *Consumption in the age of affluence: The world of food*. London: Routledge.

Fine, B., Heasman, M., & Wright, J. (1998). What we eat and why: A socioeconomic approach to standard items in food consumption. In A. Murcott (Ed.), *The nation's diet: The social science of food choice* (pp. 95–111). London: Longmans.

Fine, B., & Leopold, E. (1990). Consumerism and the industrial revolution. *Social History, 15*(2), 151–179.

Fine, B., & Leopold, E. (1993). *The world of consumption*. London: Routledge.

Fine, B., & Simister, J. (1995). Consumption durables: Exploring the order of acquisition. *Applied Economics, 27*(11), 1049–1057.

Hansen, A. (2018). Meat consumption and capitalist development: The meatification of food provision and practice in Vietnam. *Geoforum, 93*, 57–68.

Isaacs, G. (2015). *The commodification, commercialisation and financialisation of low-cost housing in South Africa*. FESSUD Working Paper Series, No. 200. Retrieved from http://fessud.eu/wp-content/uploads/2015/03/FESSUD_WP200_Commodification-Commercialisation-Financialisation-Low-Cost-Housing-in-South-Africa.pdf.

Jackson, P., Russell, P., & Ward, N. (2004). *Commodity chains and the politics of food*. Cultures of Consumption Working Paper Series, No 18. Retrieved from http://www.consume.bbk.ac.uk/publications.html.

Leslie, D., & Reimer, S. (1999). Spatializing commodity chains. *Progress in Human Geography, 23*(3), 401–420.

Lis, P. (2015a). *Financialisation of the system of provision applied to housing in Poland*. FESSUD Working Paper Series, No. 100. Retrieved from http://fessud.eu/wp-content/uploads/2015/01/2-LisP_FESSUD_WP8_housingPoland-working-paper-100.pdf.

Lis, P. (2015b). *Financialisation of the water sector in Poland*. FESSUD Working Paper Series, No. 101. Retrieved from http://fessud.eu/wp-content/uploads/2015/03/FESSUD_WP8_Financialisation_of_the_water_sectorIn-Poland_working-paper101.pdf.

Mader, P., Mertens, D., & van der Zwan, N. (Eds.). (2020). *International handbook of financialization*. London: Routledge.

Mansvelt, J. (2005). *Geographies of consumption*. London: Sage.

Mattioli, G., Roberts, C., Steinberger, J., & Brown, A. (2020). The political economy of car dependence: A systems of provision approach. *Energy Research & Social Science*, 66. https://doi.org/10.1016/j.erss.2020.101486

OECD. (2002). *Sector case studies series household energy & water consumption and waste generation: Trends, environmental impacts and policy responses*. Report by Working Party on National Environmental Policy, Organisation for Economic Cooperation and Development, Paris. Retrieved from http://www.oecd.org/officialdocuments/publicdisplaydocumentpdf/?doclanguage=en&cote=env/epoc/wpnep(2001)15/final.

Ritzer, G. (1993). *The McDonaldization of society: An investigation into the changing character of contemporary social life*. Thousand Oaks, CA: Pine Forge Press.

Ritzer, G., & Jurgenson, N. (2010). Production, consumption, prosumption the nature of capitalism in the age of the digital 'prosumer'. *Journal of Consumer Culture, 10*(1), 13–36.

Robertson, M. (2014). *Case study: Finance and housing provision in Britain*. FESSUD Working Paper Series, No. 51. Retrieved from http://fessud.eu/wp-content/uploads/2013/04/Case-Study_-Finance-and-Housing-Provision-in-Britain-working-paper-51.pdf.

Robertson, M. (2016). *The system of provision for housing in selected case study countries*. FESSUD Working Paper Series, No. 193. Retrieved from http://fessud.eu/wp-content/uploads/2015/03/FESSUD_WP193_The-System-of-Provision-for-Housing-in-Selected-Case-Study-Countries.pdf.

Robertson, M. (2020). How and why are things consumed? In E. Van Waeyenberge & K. Deane (Eds.), *Recharting the history of economic thought* (pp. 69–88). London: Routledge.

Santos, A., Serra N., & Teles, N. (2015). *Finance and housing provision in Portugal*. FESSUD Working Paper Series, No. 79. Retrieved from http://fessud.eu/wp-content/uploads/2015/01/FESSUD_Working-Paper-Series_Santos-Serra-Teles-2015-79.pdf.

Sen, A. (1981). *Poverty and famines*. Oxford: Clarendon Press.

Teles, N. (2015a). *Financialisation and neoliberalism: The case of water provision in Portugal*. FESSUD Working Paper Series, No. 102. Retrieved from http://fessud.eu/wp-content/uploads/2015/03/Financialisation-and-neoliberalism-the-case-of-water-provision-in-Portugal-working-paper-102.pdf.

Teles, N. (2015b). *Prospects for financialized water provision: Going beyond public/private ownership debates*. FESSUD Working Paper Series, No. 177. Retrieved from http://fessud.eu/wp-content/uploads/2015/03/FESSUD_WP177_Financialised-water-provision-beyond-public-private-ownership-debates.pdf.

Thurnell-Read, T. (2018). The embourgeoisement of beer: Changing practices of 'Real Ale' consumption. *Journal of Consumer Culture, 18*(4), 539–557.

Yilmaz, G., & Çelik Ö. (2015). *Finance and system of provision of water: The case of Istanbul*. FESSUD Working Paper Series, No. 153. Retrieved from http://fessud.eu/wp-content/uploads/2015/03/Water_METU_WP_FESSUD-153.pdf.

CHAPTER 2

The SoP Approach: Theoretical Background and Empirical Practice

2.1 Introduction

A SoP for a good or service is understood as the integral unity of the economic and social factors that go into both its creation and its use. As explained in Chap. 1, the SoP approach had its initial roots in the study of private consumption, with commodity production for sale as the main form of provisioning. The approach has since been adapted to encompass other forms of non-commodity provisioning including by the state, for example, through the provision of social, public or infrastructural goods such as education, health and transport, and other systems in which public and private roles intersect. Each SoP involves what might be termed a chain of provision, how do we get our food, our classroom and syllabi, our electricity supply, and so on. At one level, it is easy to specify something as a SoP, as 'systems' are heavily used in popular and scholarly parlance. We are all used to talking about the food system or the transport system, and the health and education systems, the energy system and so on, just as systems types of analysis have also become increasingly used in academic research, as with Global Value Chains and Global Commodity Chains, discussed in more detail below in this chapter. Systems-based analysis has long been part of sustainability literature (see Chap. 6).

But the SoP approach has distinctive features that set it apart from other types of systems analyses. For the SoP approach, different areas of provisioning are seen as forming separate but integral social entities that are themselves constituted out of the chains of activities that are involved,

K. Bayliss, B. Fine, *A Guide to the Systems of Provision Approach*,
https://doi.org/10.1007/978-3-030-54143-9_2

from production through to consumption or, more generally, use or application. In the context of (private) commodity provisioning, each SoP is seen as distinct from, if interacting with, others and to vary significantly from one commodity (or commodity group) to another. The operation of a SoP is shaped by multiple factors—such as the social, political, economic, geographical and historical. These in turn can give rise to distinct, commodity-specific cultures of consumption, and so inform the patterns of practices, ideas and meanings that shape consumption configurations. SoPs work through multifarious channels, from policy design and practice through production, financing and commercial operations, through sale and use (and disposal in some cases). SoPs are attached to narratives and discourses and these change over time. In principle, we should not think of one element in the system, without connecting and relating it to others in the chain of structured activities, which can reach across local, national, regional, and global sites and influences.

As a result, the SoP approach examines the system as a whole even if the focus of interest, for research or policy purposes, might be on one aspect of the SoP alone. Thus, the classroom or the hospital is not to be seen in isolation from other aspects that underpin it—for example, financing, supplies, legal frameworks, institutional arrangements, curricula design, personnel, motivations of families around education and heath, opportunities for access, and differential needs depending on contexts. Also important is the set of cultures attached to provisioning and consuming. The SoP approach is not only interested in how the *consumer* understands what is being consumed, and the processes leading to that consumption, but also how do other agents understand the process whether in production, regulation, finance or otherwise. Moreover, each SoP is embedded in a corresponding set of broader determinants, reflecting socioeconomic and sociocultural factors, as well as how these play out in specific contexts—the supermarket as access to food, the hospital to health care, the mortgage to housing, as well as gender and advertising as influences on what is consumed, how and with what significance, and so on. Accordingly, each SoP, whether by sector or location, will have its own unique character, even though it might share some common factors with other SoPs.

For the SoP approach, what is relevant is not just the chain from production that goes into providing a commodity or service but also the way in which it is provided to different classes of consumer. If we consider the fashion system, there are clearly differences in the degree, forms and content, for example, in how clothing is designed, produced and sold across gender, class and race, the young and old, and further on in terms of what is fashionable, and how (quickly) it changes over time and across locations, although globalization may be narrowing, as well as creating particular, differentiations. The fashion system indicates that, like other such systems, it is complex across a number of dimensions—how it delivers, to whom, under what conditions or, to put it another way, the SoP approach is concerned with who gets what, and how, in which there can be profound differences in (in)equalities, for example, of opportunities, access and quality.

Given these broad domains of concern to be covered, some structure of analysis is needed, and this is the focus of this chapter. There is clearly a danger that the approach could be so broad as to become meaningless in the absence of some systematic analytical framing. The chapter proceeds with a section on the theoretical influences behind the SoP approach. This is followed by a discussion of the five fundamental components of which a SoP is constituted and then considers more practical methodological concerns in applying the approach in practice. The SoP approach is compared with other 'systems' approaches before the final section concludes.

2.2 Theoretical Underpinnings

As stated previously in Chap. 1, the impetus for deriving the SoP approach came from dissatisfaction with the fragmented and siloed approaches of the social sciences to consumption more broadly. Hence the approach embraces many of the contributions from sociology and anthropology and other disciplines. However, the SoP approach considers that treating these as elements in an integrated vertical chain of provisioning offers a more promising way of understanding consumption outcomes. The SoP approach has been presented and was deliberately designed to be of appeal as a malleable synthesis across diverse approaches and hence is more or less open, if not neutral, in methodological and theoretical terms.

However, in devising this framework, the SoP approach has clear theoretical origins and influences (as set out retrospectively in Fine 2013) in view of its initial goals, context and practitioners. Specifically, with focus upon consumption through commodity provision, Marxist political economy (MPE) provides an opportune starting point, precisely because of its emphasis upon commodity production as a structured process, involving (class) agencies and corresponding relations of capitalist production in provisioning. This is not to reduce consumption to the imperatives of production but, rather, to set the context within which the structures, processes, relations and agencies of consumption do themselves prevail, something that has tended to be overlooked by MPE itself as well as by political economy more generally.[1] The point is to address, to engage positively with, consumption in its own right, not as a passive reflection of production but neither independently of it.

From its beginnings, then, the SoP approach focused upon consumption derived from private commodity production and, as a result, was heavily influenced by the theory of commodity production to which it is generally acknowledged, even if often critically, that Marx himself, and MPE are major contributors. In the corresponding literature, the theory of commodity production is heavily oriented around what is known as value theory. This is traditionally concerned with two major questions—what is the *source* of value in a commodity, can it, for example, be confined to exploited wage labour; and what is the *relationship* between value as labour time and the formation of prices. These questions have profound implications for understanding the development of capitalist production and the accumulation of capital, and there has been much controversy surrounding the relationships between value, price and profit.[2]

For the SoP approach, what is of particular significance is the way that value is conceived in MPE. The approach draws on what might be termed some of value theory's qualitative insights into commodity production. Specifically, in a capitalist society, the means of production (capitalist firms) are generally privately owned and their products are distributed through market exchange. Because products are exchanged through the market in what are more or less impersonal relations, they appear as commodities primarily disassociated by their purchasers/consumers from the production processes (and other processes) by which they have been brought to the market. And, in particular for MPE, it is the value relations derived from production that are not apparent in these market transactions. Products might just as well have fallen out of the ether into the hands of those who are selling them as is the case, for example, in the simplest models of mainstream economics relying on exchange alone, resulting from supply and demand, in the absence of any production. For Marx then, the overwhelming attention in daily life, and indeed in what he termed vulgar economics for this reason, to the commodity form (buying and selling on the market) in capitalist society, fails to reveal directly, or indeed conceals, the social relations, structures, agencies and processes by which those commodities have been delivered. For example, look at a chair and its price and you cannot see a carpenter and how the chair has been produced in and of itself nor the materials and technologies used. Furthermore, it is not revealed why it is worth more or less than the work that has gone into producing something else such as an item of clothing. Similarly, the online shopper is mostly oblivious to the offshore financial practices of the service provider or the terms and conditions of zero hours contract of the delivery driver. This contextual backstory is typically unknown in capitalist exchange even though we know the relative 'value' of each of the products (or absolute 'value' in money) by virtue of their mutual exchangeability as indicated by prices.

All that commodities reveal to you in a market exchange, then, are their (relative) prices, against which you gauge their use to you and whether you are able and willing to buy them or not, for whatever purpose. The life of the commodity prior to sale, in terms of the materials and technology used as well as the labour process, the social and material environments and the broader character and context of the chain of provision, is at most marginally revealed, if at all, in the acts of sale, purchase and use of the commodity. The idea that labouring to produce things is not apparent in commodities as they are exchanged on the market as opposed to their relative prices is what Marx famously termed 'commodity fetishism'. The processes of 'fetishisation' mean that the products of individual producers become exchangeable in a market organized through competition and the social relations that underpin production become invisible other than to those directly engaged in them (Hartmann 2014).

However, for MPE, the concern is that, while the commodity does indeed have a price (or an exchange value) and can be bought and sold and used, this is only because underpinning those price relations are (concealed) social relations of production. Moreover, those concealed social relations are dependent not just on labour, but potentially on exploited wage labour. For MPE a commodity embodies different processes that both create a surplus as the source of profit but also conceal its exploitative character through the apparent freedom of exchange between capital and labour on the market.

Beneath the surface of an apparently free-standing commodity, then, sits a complex web of interactions. Value theory in production is concerned not just with the labour time of production but with the *relations* that *agencies* set between themselves in the *processes* of provisioning commodities for the purposes of consumption, and how these are *structured* across production (for profit) through exchange. Value is associated with, often and inevitably contested, relations of production, for example, with the capitalist typically pushing to extract as much surplus from the production process as possible, and outcomes from this will be governed by power relations at and around the work process more generally. Over and above this, processes of exchange and competition are far from harmonious and also involve inherent tensions across consumers and producers in the distribution of value which emerges in the pricing process.

Surplus value accrues not just as profit but also as rent and interest payments, depending on how land and finance are attached to commodity relations. In the more complex production processes of a modern economy, tensions may emerge regarding the distribution of value across other groups of agents involving, for example, financiers and regulators. Much of this is also concealed, or even falsified, if the focus is primarily on market relations alone as if, for example, land can itself produce value simply because it can command a rent,[3] or a corrupt official produces value because in receipt of a bribe. Yet these outcomes emerge from the specific conditions in which they are situated.

To understand consumption then requires an understanding of the processes by which the consumed was created. Consumption items do not come from thin air and the act of consuming is to participate in a potentially contested chain of activities connected to production. In order to understand consumption, both quantitatively and qualitatively, we have to delve beyond the exchange relations by which individual consumption is proximately derived, and penetrate into what goes on beforehand in the (concealed, possibly misrepresented) activities that precede exchange and consumption. The SoP approach builds on this MPE interrogation of the commodity and the social relations that underpin production not just because of their social implications but also because the production processes underpin and interact with what is consumed and how.

Also significant for the SoP approach is *why* consumption takes the form that it does. At an apparently mundane level, for example, how and in what form does the McDonald hamburger become iconic (Fine 2007)? At the other extreme, what is it that makes for the commodification of works of art and is this at the expense of something to be dubbed artistic integrity (for a recent discussion see Haiven (2018))? In contrast, at the heart of mainstream economics, not only use values but also the consumers themselves are effectively fixed—all you need to know is utility, income and prices and the consumer optimizes accordingly subject to equally prespecified conditions of supply.

But from where do such values come and why might discrepancies occur between a product's exchange value (its price) and use value (its ability to satisfy needs on the basis of its material properties)? On one level there is a kind of functionality at work. A car has a clear use value in terms of transportation. Yet why would one car be preferred over another? Exchange value (price) may be an issue. But there are more factors at work. Why use a car over public transport? An item of clothing might keep you warm but why pay a premium for a particular designer label? Clearly advertising plays a role but cultural associations run deeper and wider, for example, in association with whether provisioning is public or private, or environmentally friendly or not, as in use of plastic (such as the shift from tap to bottled water, and then to the market for water bottles). Media coverage and policy are significant but still there are questions over the origins of cultures of consumption let alone whether they are worked upon successfully or not by advertising or other messages. How and why is it that particular cultures become attached to specific goods and services? The associations between how things are provided and the mutually conditioning cultural systems to which they are attached are core concerns for the SoP approach (Fine 2002 and see Chap. 3).

For the consumer, in the same way that commodity fetishism is associated with concealed labour relations, use value comprises a wealth of factors which are not immediately obvious, including the material practices of provision as well as the way these are culturally represented (or under- or even mis-represented) and received, and reflected upon, as such in the practices surrounding consumption. Here, the SoP approach draws upon the work of Wolfgang Haug (1986). Haug suggests that there is an underlying tendency of capitalist production to be subject to degradation of use values (in pursuit of lower costs) and that this is compensated for, in order to sustain sales, by endowing commodities with meanings such as a sexualized content, not least through advertising. This gap between actual and presented use values is what he terms the 'aesthetic illusion'. For Haug the pursuit of profitability, means that the way the commodity is produced (through, possibly successfully concealed, degradation) has influence upon how it is (presented to be) conceived.

Haug's work provided valuable theoretical insights for examining the dynamic social reconstruction of use value through emphasis upon the manipulation of cultures to induce purchased consumption. There is a potential diversity between the commodity as a use value bound for consumption and as an exchange value deriving from capitalist production.

However, the SoP approach takes aesthetic illusion as a critical point of departure in three ways. First the transformation of commodities is not confined to degradation, and production quality may improve rather than decline in the process of capitalist accumulation (e.g., through changes in production processes, technological advances, improvements in quality of inputs). Second, the putative illusion is affected by a number of components in the chain of provision aside from production and retailing including transport and storage—and corresponding freshness—for example, quite apart from design and advertising in the very broadest sense. Third, there is no account taken of the consumer's role in how the gap in the aesthetic illusion—commodity as it is, relative to the commodity as it is perceived—is generated, reproduced and/or transformed and bridged. For, when the consumer is considered, the aesthetic illusion can itself be seen to be illusory as the distinction between actual and perceived use values dissolves into investigating the meanings of consumption (rather than the gap between an old and a new illusion). This can better be done directly, rather than through how it was previously and subsequently understood (and experienced) in light of change in provisioning, if not setting history aside. While intuitively plausible, there is no practical way of understanding the aesthetic illusion without knowing the starting point of the two sides of the gap and how these themselves shift. So the commodity as is and as understood might just as well be addressed directly without reference to the aesthetic illusion, otherwise needing to engage in an infinite regress to how the commodity was (falsely) conceived through history to the present. The active role of the consumer in negotiating an understanding of the consumed in light of what it is, and especially how it is accessed and presented, is generally known as 'reflexivity' and discussed below.

But the question remains as to how such cultures should be understood. While the aesthetic illusion gives some guidance, albeit limited, to the multiplicity of factors that lead to the meanings attached to the consumed by the consumer, the cultural content of the commodity is not only related to the material system of provision but also to wider cultural influences such as gender, class, ethnicity and national identity. Furthermore, Haug fails to take account of how products attain their actual qualities, and perceived qualities, in the first place as the basis from which they move in relation to one another (as more or less degraded products and how they are represented). How do all of these elements fit together and relate to one another? One answer has been to appeal to the notion of a 'circuit' of culture, originally deriving from du Gay et al. (1997, and see Chap. 1). This incorporates the important insight that the origins of influence on the cultural content of commodities are multiple, and derive from each and every aspect connected to the provisioning of the commodity (and not just advertising and the shifting material nature of the commodity whether through degradation or otherwise). But it is not clear why the metaphor of a circuit is justified, other than movement across time and place in going from production to consumption, possibly over and over again. Does culture come and go, or simply return for reworking at a later date having been transformed by others at various points along or around the chain of provision, a sort of game of Chinese whispers?

The SoP approach is, then, situated at the centre of these diverse threads around the nature of value and linking production to consumption. Commodities are considered to be constituted of, often contested, social relations in the production process. In consumption, use values are open to a multiplicity of influences including being shaped by material cultures. The SoP approach is concerned with the intersections of these, and to consider the ways in which contextual factors, beyond production and consumption, influence use and exchange values. Taking gender, for example, the study of consumption in, say, the fashion industry typically focuses on the gendering of the clothes purchased or, possibly, displayed on the catwalk. But what of the gendering in the manufacturing of the clothes even if attention is paid to the design of haute couture, and what roles do these play in relation to one another and the reproduction of the gendered fashion systems (different ones for males) and corresponding consumption as a whole. Often cheap, or even expensive, fashion relies on cheap labour by some (possibly women) and generates bloated rewards for designers, models and (some) retailers (see Brooks 2015 and Chap. 4).

Rather than a circuit of culture as organizing principle, the SoP approach proposes that each SoP is also attached to its own integral and corresponding cultural *system*. The cultural system derives content from each and every material aspect of the SoP, as proposed by du Gay et al. (1997), although it does so in ways that are not rigidly predetermined. This is to move from the material structures and processes that condition the cultural system attached to commodities, to the nature and origins of that culture itself. In part, this is because consumers are not passive recipients of the culture attached to commodities (or anything else for that matter) but are what has previously been termed reflexive. But they are not reflexive in circumstances chosen by themselves.

The most immediate, if not exclusive, external determinants of that reflexivity are to be found within the components of the SoP itself. But, clearly, the SoP cannot dictate how consumers reflect upon what they consume although its agents are usually inclined to exercise an influence, for example, through advertising in pursuit of successful sales. Further, consumers engage in a variety of practices around consumption that are not necessarily reducible to the act of consuming itself, as in family meals, display, emulation and distinction, and so on (Warde 2005). For the SoP approach the consumer is located within an extensive system ranging from the social relations of production to the material cultures of consumption. It is the interaction of these complex components, dependent on the context and the commodity in question that lead to specific patterns and meanings, or cultures, of consumption and which need to be unpacked.

2.3 Fundamentals for a SoP Analysis

While the SoP approach is open to a large degree with regard to methodologies (see below), SoP research necessarily incorporates certain core elements which have long been used widely across the social sciences.[4] Significantly, though, these are far from common in mainstream economics as such, which is dominated by reductionism to a market/non-market dualism based on individual optimization on the basis of given utility functions.

This section sets out the fundamentals on which SoP research can typically begin to be built. The categories deployed are both general and overlapping and may feature to varying degrees in specific research depending on the area and purposes of investigation. They may be grand and global or highly specific and local in content.

1. Agents/agencies—SoPs are determined by the participants in the provisioning system. Incorporated are those who produce and those who consume but also wider bodies such as trade unions, consumer groups, the state and those who affect delivery of finance, investment, technology, and so on. Within these categories will be subcategories. Consumers will fall into different strata possibly delineated by income or by geography or gender or race or by use of elements of a service or product, with providers distinguishing between business and residential customers, for example. There may be institutional representation of interests whether through the state

or otherwise, as with consumer representation groups. Producers will take different forms with distinctions across local and global reach and networks. Within the category of producer also lie subgroups, for example, between worker, management, shareholder and financier. The state will incorporate different facets, possibly as provider, regulator, policy maker or even consumer itself with distinct entities operating at national and local, and possibly international, levels. Global institutions such as the World Bank, the IMF, ILO, EU and WHO will have a role to play in some cases. Each of the agents specified in the SoP has an impact in defining the way in which outcomes emerge, and each will engage in the SoP differently, often with competing objectives and priorities. Each will also have its own perspective on the SoP including objectives and methods for achieving outcomes, for themselves and/or others, and how these should be projected and influenced. The level of disaggregation or detail of analysis will depend on research questions and purposes.

2. Structures—the agents in the SoP will be operating within historically and socially specific forms of provisioning that evolve over time. Such structures take different forms and, for example, may be organizational, institutional and social, formal or informal. At the grand theoretical level, there is the structuring of capitalism itself in terms of production and exchange and for profit or not, often termed economic reproduction. Economic reproduction is embedded within broader social structures and social reproduction. Consumption straddles all of these structures to a greater or lesser degree.[5] But, at the level of specific SoPs, in time, place and sector, such broad structures are more concrete, although they may be complex with extensive potential scope. These will involve the vertical structures of provisioning themselves (across not only production but also financing, marketing, state regulation, and so on). These specific vertical structures will interact with horizontal factors. Important in consumption, for example, will be levels of income, and all the structured variables of economic and social status such as race and gender.

   This points to the potential importance of inductively identifying context-specific factors in which SoPs are located, varying over organizations, legal frameworks, the relative weight of public and private sectors, as well as how power and hierarchies are structured.

Analytically, this runs the risk of empirically identifying causal factors (the structures) and then using these to explain outcomes by those structures—women suffer from sexism, etc. A deeper level of causal determination can, however, be pursued within the SoP approach by uncovering the ways in which different structures are reproduced, or transformed, as a result of the integral nature of SoPs across their various elements, both through the agencies involved and how they go about their business. So, if we take gender bias, the SoP approach needs to explore how this is manifest in different forms across different segments of the SoP and why and with what effects. Alternatively in the housing sector the institutional structures such as different forms of housing tenure (owner-occupied, private rental, social housing) intersect with social structures to lead to social outcomes which might take the form of inequalities in levels of, and access to, housing.

3. Processes—each SoP is shaped not only by *who* does what and in which structures but *how* it is done as a structured sequence of activities, something that goes far beyond matching supply with demand. Again, processes can be understood both in systemic and abstract terms such as globalization and privatization, as well as in specific activities within the SoP (the labour process, advertising, and so on) and around the SoP (gendering, etc). Processes are engaged in everyday life but these are much more broadly situated in social processes. This is so not least for privatization, for example, as it changes both the structures and agencies in provisioning as well as how the consumer engages with provisioning both in terms of how consumption is accessed but also how it is perceived, see below and next chapter on cultures. Especially important is the contrast between welfare as opposed to market provisioning as it is marked by many different aspects that go beyond payment (or not) for what is provided. Other relevant issues would include levels of entitlement in principle and practice and how these are managed. Inevitably, such processes are subject to tensions and conflicts. The SoP approach in practice most recently has laid considerable emphasis upon globalization, neoliberalism and financialization as influential processes with clear implications for consumption, although each of these is highly contested in terms of meaning and content across the literature, and how they give rise to variegated outcomes. Some studies are discussed in more detail in Chap. 4.

4. Relations—SoPs are constituted upon and, in turn, constitute and reproduce relations of class, gender, race, caste, etc, contingent upon who exercises power, and how, and with what purpose (and meaning to participants) and, as such, are open to contestation and conflict. So the relations upon which SoPs are founded are differentiated by the roles of capital (or state as employer) and labour in production and other commercial (or non-commercial) operations through to the relational norms and implicit codes of practice that are attached to levels and meanings of consumption. The notion of norms in relation to consumption is discussed in Chap. 3. For the SoP approach, there is a need to specify relations across the three headings above: how are relations structured; how are positions in those structures occupied by different agencies; and what are the processes by which the SoP is reproduced or transformed.
5. Material cultures—each SoP will also be characterized by the different meanings of provisioning for those who are involved with it in whatever way, whether by providers, citizens, consumers, policymakers, the media or scholars. A crucial part of the SoP approach is to emphasize that there is an integral relation between each SoP and the cultures that are attached to or, more accurately, interact with it. Once again, more generally, study of the relationship between the material and cultural worlds is highly controversial and, as will be laid out in the next Chapter, the SoP approach seeks to finesse these controversies in the ways in which it frames material culture as a product of, and contributing factor to, SoPs themselves.

The rationale for investigating SoPs through agencies, structures, processes, relations and cultures is to ensure that both sufficiently holistic a view can be taken and that this can allow for specific SoPs to be addressed and set in context. A SoP, then, might initially appear to be reducible to a simple matter of more or less structured supply and demand within a particular sector but, as soon as we begin to investigate what is provided, how and to whom, and with what meanings to its participants, it becomes a matter of unravelling a whole series of complex but interconnected issues.

Context and specificity are key to SoP research. As Southerton (2011, p. 1424) writes of the SoP approach: ‘it would not offer explanations about the consumption of peanut butter or blueberries based on theories that have been developed to explain the consumption of motor vehicles’. Each sector will have its own SoP and within sectors there may be distinct smaller SoPs, for example, there may be SoPs for agriculture, for dairy and for cheese. There are also characteristics of the providers and the consumers which affect the ways in which they engage with the SoP such as age, gender, location and ethnicity. To reiterate, the SoP for cars is different from that for houses and different again from fast food and fashion, and these will be different from country to country, from urban to rural and today from a century ago.

In principle, then, each SoP needs to be addressed by reference to the material and cultural specificities that take full account of the whole chain of activity, bringing together production, distribution (and access), and the nature and influence of the conditions under which these occur. Even at the level of empirical narrative, this leaves open some degree of ambiguity and choice on the part of the researcher. In part, this is because of the need to identify the scope of specific individual SoPs themselves. Thus, for example, private and public housing will likely be separate but connected SoPs, as may be the case with private rented and owner-occupation, although each will share some of their elements in common. Similarly, bottled and piped water will almost certainly be perceived as belonging to separate, if overlapping, SoPs. In addition, even if the SoP itself, and its elements, has been empirically identified, possibly uncontroversially, it is still open to be understood in very different ways both within and across disciplines, conceptualizations, methods and theories.

2.4 The SoP Approach: Challenges in Practice

SoP analyses require taking into account an extensive range of factors which can be daunting for the researcher facing limitations both in terms of resources and access to information. Clearly not every element of each component in the chain of provisioning as well as every contextual or horizontal factor can be thoroughly investigated. How then should a SoP study be conducted in practice? Chaps. 4 and 5 provide some examples of the application of the SoP approach, and some more general methodological pointers are provided below.

Unravelling a SoP may require a mix of methods including data analysis and quantitative structured interviews as relevant to the area under investigation. But it does have some minimum requirements that are, as it were, non-negotiable. These are the features of openness and flexibility which, even if based on particular foundations, allow the approach to draw both constructively and critically on other approaches, across academic disciplines, in light of how other contributions offer theoretical or empirical insights into how the SoP functions without necessarily being sufficiently comprehensive or in conformity with the SoP approach's foundations. While the SoP approach is methodologically and theoretically open to a considerable degree, this does not mean that it is analytically empty let alone neutral (see above). Indeed, it definitely rejects other approaches, where they are inconsistent with the SoP approach's open and yet contextually situated stance, particularly the deterministic assumptions regarding the optimizing behaviour of individuals characteristic of mainstream economics. Investigations of more specific, narrower, segments of production and consumption studies can be incorporated into the wider SoP analysis. This might, though, involve transformation in the understanding of these elements in and of themselves and by virtue of locating them more broadly within the SoP approach. The commodity might appear differently to the consumer if known to be produced by child labour, by spoiling the environment and so on. With advertising many may be able to distinguish between its role in selling and in informing, without necessarily finessing the balance between the two in constructing the cultures attached to the consumer and the consumed. The effects are more profound, otherwise, why, if we understand that most advertising is misleading, literally fantastic in many instances, does it persist? The SoP approach is also open to contribute to the understanding of normative issues and the role of ideational factors, as well as seeking to understand why and how they are deployed, and with what effects.

The approach is heavily inductive, in contrast to mainstream approaches within economics which tend to apply a deductive approach with hypotheses based on a world view that is then tested and accepted or rejected. The SoP approach is more iterative. For the SoP approach, there is not necessarily an initial research question or hypothesis to be tested but rather an identifiable area of investigation. As a result, identifying a SoP is not a matter of taking an ideal type off the shelf that is deemed to fit best, nor does it involve applying a predetermined template. Rather, it is a reasoned dialogue between theory and evidence in light of purpose, ranging over the chain of provision, its broader determinants, its material culture and its context.

A challenge for SoP studies is setting boundaries. It is not necessarily clear how a SoP should be delineated as, at a holistic level, everything is related to everything else particularly as broader factors such as income, gender and race tend to span all SoPs. Why not one big social SoP, or should it be thousands of mini-SoPs at the other extreme? How should we approach systems within systems such as the SoPs for milk, dairy and food? And what about home meals and eating out, each of which straddle food SoPs and other activities and determinants, not least burgeoning home deliveries? The way a SoP is identified depends, to some extent at least, on the question at hand and area under investigation. For research purposes it is usually necessary to shine a spotlight on the elements of the SoP that are of particular relevance to the issue under consideration. Ball's structures of provision approach to housing, which was a strong influence on the development of the SoP approach (see Chap. 1), provides a useful example. Ball argued that researchers interested in the incidence and impact of state subsidies on housing outcomes, especially the distribution of quality of housing, needed to take into account considerations beyond balance of tenure types because the way that housing is provided determines the characteristics of different tenures. Housing is discussed in more detail in Chap. 4.

The detailed methodology and overall line of enquiry will, thus, depend on the specific area under investigation in dialogue with the empirics of the SoP itself. As already indicated, the SoP approach is based upon empirically or inductively identifying integral systems of material and cultural factors that mutually condition one another. By their very nature, the boundaries of SoPs have to be identified (and, as indicated, according to purpose) in the course of the research process and cannot be analytically predetermined (avoid off-the-shelf templates as also previously indicated). The question of SoP boundaries can, thus, only be answered contextually through identification of the integral structures themselves in practice (with a presumption towards widening rather than narrowing the factors to be taken into consideration).

2.5 The SoP Approach Compared to GCC Approaches

Systems thinking around commodity consumption, although far from novel, is increasingly attracting attention across disciplines and topics from medicine to education as analysts are seeking to apply a more holistic approach to understanding how constituent parts interrelate to achieve specific outcomes that might be affected by targeted interventions. Food systems approaches, for example, connect food and hence farming and food processing with public health, not least given the diseases of affluence associated with shifting diets underpinned by rising incomes, see Chap. 5. Campaigns for environmental sustainability have long been linked to systems-based analysis (Meadows 2008) and Chap. 6.

In this light, whilst the relationship between the SoP approach and systemic analyses is taken up in Chap. 6 in broader detail, it is useful, in order to bring out its distinctiveness, to contrast the SoP approach with that based on global commodity chains (GCC) and global value chains (GVC).[6] In principle, there would appear to be a close affinity between the GCC/GVC and SoP approaches. Each emphasizes attention to the chain of activities connecting production to consumption, and how such a chain offers an integral form of commodity supply, with a corresponding structure and dynamic and power relations between agents (primarily, for GCC/GVC at least initially, oriented towards explaining who appropriates value along the chain, how much and how). But each approach has also evolved from different origins and with different analytical trajectories, with little or no dialogue between them let alone convergence. Like many other systemic approaches such as GVC analysis, the SoP approach is dedicated to taking a holistic view of the chain of, and from, production that underpins consumption patterns. However, SoP departs from the GVC approach in its understanding of the factors that shape consumption, arguing that these are imbued with case- and context- specific meanings, cultures and material features. Thus Fridell (2018), for example, points to the way in which the GVC approach focuses upon power relations along the chains of provision (to appropriate the value generated) but neglects the power relations that underpin these from outside of the chains of provision.[7] These will vary across sector, location and over time and will relate to the agencies under examination.

GCC/GVC began with an attempt to situate sector-specific production within World-Systems Theory. It did so by suggesting ideal-types of chains, initially buyer- and retailer-driven, highlighting an abiding concern with (the material culture of) power and governance along such chains, as avenues for the appropriation of value added.[8] Inevitably, this two-fold classification proved inadequate for addressing the empirical diversity of chains themselves, and the typology was extended accordingly. Thus, GCC/GVC research has been driven by an evolving (and expanding) classificatory scheme to accommodate the burgeoning case study evidence that otherwise fails to fit—and ultimately leading to reliance upon global production networks (GPN) as a way of chasing increasing empirical diversity in the fragmented but global organization of production.[9]

The commitment to empirical case study is a strong and driving factor in GCC/GVC research—follow the chain. Its counterpart is a corresponding weakness in theory, in terms of depth, consistency and rationale. No doubt its practitioners would strenuously disagree but they seem to have been subject to a syndrome of off-the-shelf appropriation of whatever theoretical fragments suit shifting empirical needs or intellectual fashions, implying that GCC/GVC has, consequently, had little to offer in return to the social theories on which it has drawn. Indeed, tongue in cheek, it is tempting to see the GCC/GVC approach as a reflection in its own mirror, imposing theoretical governance over its unruly case studies by a chaotic assemblage of concepts.[10]

A particularly acute example of the lack of such broader self-reflection on the part of the GCC/GVC approach is its failure to examine what defines and determines the chains themselves, with this generally taken as self-evident by virtue of case studies. This follows from an almost exclusive preoccupation with vertical (follow-the-chain) analysis, to which is added the coordination/governance of the given chain, confined exclusively to private consumption. But what of horizontal factors, both within and beyond the chains? Thus, not only no chains for steel, energy, transport and finance and so on, let alone health, education and welfare, but these also potentially fracture the integral nature of the chains that can be identified. For finance or, more exactly, financialization, for example, the rise of equity finance and of futures markets for commodities not only brings into question the functioning but also the definition of chains. Is it possible that along the chain (governance) factors may have become subordinate to across-chain financialization?[11] So GCC/GVC analyses do not neglect horizontal factors, but the selection of those that are included is limited in both scope and depth, as is the scope of sectoral examples provided by the approach.

As a result, it is hardly surprising that consumption should have remained off the GCC/GVC/GPN agenda for so long despite the heavy influence of the post-Fordist flec-spec/filière approach. For, whilst post-Fordism emphasized small-batch production serving consumption in niche markets, it moved no further than this, certainly relative to the explosion of consumption studies inspired by postmodernism. And, had such a cultural turn in consumption been adopted by the GCC/GVC/GPN approach, it would have swept away and not just been subordinated to what would have been perceived to be the economistic and reductionist attachment to ideal-type chains. Such antipathy across the social sciences to the GCC/GVC/GPN approach, more by way of benign neglect than active critique, has been reinforced by two further factors. First, the scope of the approach has been extremely limited by the nature and number of the case studies, centred on a few sectors globally organized for final consumption. Despite its origins in World-Systems Theory, examination of the implications of the chains for the world system in the age of globalization and neoliberalism has been notable for their absence, both directly and in their implications for other sectors, from raw materials through to finance, e.g. those not destined for final consumption, and the role of the state. As Bernstein and Campling (2006a, p. 240) put it:[12]

> the commodities on which current 'commodity studies' concentrate are above all those in the realms of personal consumption. In short, there is little interest in capital or producers' goods and intermediate goods, which are no less commodities than 'exotic' fruits air-freighted from the tropics to Northern supermarkets, branded coffees or clothing, or eco-tourism.

Fine (2002) amplified this point in relation to the wider domains of public systems of (state) provision in transport, energy, housing, and so on, major sectors apparently falling outside the scope of GCC/GVC analysis, even though increasingly subject to private and global provisioning.

In contrast to the GCC/GVC/GPN approaches, the SoP approach originated with the study of consumption itself and, in this sense, traverses an analytical route in the opposite direction albeit very much more quickly if not instantaneously (in terms of returning to the issues of the material cultures of provisioning). As mentioned in Chap. 1, Fine and Leopold (1993) argued that theories of consumption had developed in a horizontal fashion within disciplines and across variables (utility maximization for economics, emulation and distinction for sociology, deconstruction of meaning for cultural studies and so on) and across commodities. Further, to integrate the various analyses together and the corresponding factors involved requires attention to specific commodity chains designated for consumption. In other words, the SoP approach sought both to identify complex patterns of consumption and to link them to systems of provisioning. And, unlike the GCC/GVC/GPN approach, it rejected ideal-types to specify chains. Rather, the integral nature of individual SoPs, global or otherwise, is seen as contingent upon the socially and historically specific forms taken by the accumulation of capital, and corresponding forms of social reproduction on which they are based and which give them their structured dynamics.

In short, in a specific study of the Star Anise commodity chain, as a "flexible" crop, Turner, Derks & Hạnh (2019, p. 278) report:

> To date, the main conceptual approaches to commodity chains include global commodity chain analysis …, global value chains …, commodity networks …, commodity circuits … and the systems of provision approach. We draw on a systems of provision approach here, as it emphasises interactions between production and consumption, focusing on the ways actors assign meaning to goods at each stage along the chain (Fine and Leopold 1993). This enables a critical look at on-the-ground connections among actors, and the operation of specific chains at a range of scales ... As such, the systems of provision approach has been argued to engage more deeply with cultural dimensions of production and consumption than other conceptual entry points, especially global value chains.

This neatly highlights the virtues of the SoP approach as well as its relations to other approaches, not least for food (see also Chap. 5), and more generally (see also Chap. 6).

2.6 Conclusion

This chapter shows that the SoP approach presents an ambitious framework for understanding the nature of consumption. As shown, the approach aims to go beyond other commodity systems type approaches to consider not just the elements of the system but their origins and the underlying power relations and conflicts which shape outcomes. The approach has not been without criticism. For example, the originators of (global) food system analyses—ironically, a source of inspiration for SoP food studies themselves—underwent a crisis of self-belief in the early 1990s in response to increasing diversity in national agricultural systems and closer attention to the corresponding empirical evidence. Temporarily at least, the proponents of food systems were induced to become critical of the SoP approach because of the problems with *global* food systems as too homogenizing an approach. But they threw out the baby of food systems with the bath water of the global. In doing so, it was mistakenly overlooked that such heterogeneity in food systems at national or other levels is precisely what the SoP approach is designed to address, see Fine (1994a, b).

In short, working across the social (and physical) sciences, the SoP approach provides a framework for conducting a comprehensive assessment of the factors that shape consumption, rooted in the real world and embracing the complexities and diversities that generate outcomes. This is not to deny that consumers themselves have agency through practice and reflexivity but this is situated within an extensive range of factors beyond the immediate reach of consumers themselves. The role of such material cultures in shaping consumption is investigated further in the following chapter.

Notes

1. There are, of course, notable exceptions such as Thorstein Veblen and conspicuous consumption as well as critical stances on consumer sovereignty from a variety of perspectives that need to be taken into account (Robertson 2020).
2. This section provides a short and select overview of certain aspects of MPE. For broader coverage, see Fine and Saad-Filho (2016) and Fine and Saad-Filho (Eds.) (2012).
3. Whether value is exclusively a social product of labour or can be a natural product is heavily debated but clearly depends upon how value is specified analytically as opposed to generally as being of worth. For an interesting discussion, see Kallis and Swyngedouw (2018).
4. For a mainstream presentation, see Giddens (1979) and Fine and Milonakis (2009) for discussion in a political economy context.
5. For this in the context of SoPs, see Fine (2020).
6. See Bair (2005) and corresponding special issue of *Economy and Society*, Gibbon, Bair & Ponte (2008) and, for critique, Bernstein and Campling (2006a, b).
7. In part, Fridell's account, which is itself highly supportive of the SoP approach, derives this insight from a concern with consumer politics from which an intra-chain, as opposed to an extra-chain, perspective (of value transfer at expense of poor producers) is at most one issue of contestation not least given concerns, for example, over climate change. See also Fridell and Walker (2019) for the way in which GCC/GVC analysis has increasingly become a fix and a fantasy for experts advising on economic and social upgrading as opposed to a more deeply-rooted critical exposition and stance.
8. This underpins fair-trading campaigns to shift value appropriation towards producers and away from retailers. See FTEPR (2014).
9. The shift from GCC/GVC to GPN derives from Sturgeon (2000). See Fridell (2018) for an account and continuing problems.

10. From the abstract of Gereffi, Humphrey & Sturgeon (2005, p. 78), we get a clear idea of this: 'This article builds a theoretical framework to help explain governance patterns in global value chains. It draws on three streams of literature—transaction costs economics, production networks, and technological capability and firm-level learning—to identify three variables that play a large role in determining how global value chains are governed and change. These are: (1) the complexity of transactions, (2) the ability to codify transactions, and (3) the capabilities in the supply base. The theory generates five types of global value chain governance—hierarchy, captive, relational, modular and market—which range from high to low levels of explicit coordination and power asymmetry'. The GCC/GVC approach has also increasingly become policy-oriented, from World-Systems to World Bank as it were (see Gereffi and Frederick 2010).
11. As late as Gereffi (2014), there is no effective reference to finance, let alone financialization, even in attention to the restructuring of chains/networks in the wake of the global financial crisis, although there is some welcome emphasis on increasing concentration of corporate ownership and control.
12. This limitation of the GCC/GVC approach was strong if implicit in the stand-off in policy analysis and debate for South African industrial policy between Kaplinsky and others in ISP (1995) and Fine and Rustomjee (1997) and Fine (1995).

References

Bair, J. (2005). Global capitalism and commodity chains: Looking back, going forward. *Competition and Change, 9*(2), 153–180.

Bernstein, H., & Campling, L. (2006a). Commodity studies and commodity fetishism I: Trading down'. *Journal of Agrarian Change, 6*(2), 239–264.

Bernstein, H., & Campling, L. (2006b). Commodity studies and commodity fetishism II: 'Profits with principles'? *Journal of Agrarian Change, 6*(3), 414–447.

Brooks, A. (2015). *Clothing poverty: The hidden world of fast fashion and second-hand clothes.* London: Zed Books.

Du Gay, P., Hall, S., Janes, L., Mackay, H., & Negus, K. (1997). *Doing cultural studies: The story of the Sony Walkman.* London: Sage.

Fine, B. (1994a). Towards a political economy of food: A response to my critics. *Review of International Political Economy, 1*(3), 579–586.

Fine, B. (1994b). Towards a political economy of food. *Review of International Political Economy, 1*(3), 519–545.

Fine, B. (1995). Flexible production and flexible theory: The case of South Africa. *Geoforum, 26*(2), 107–119.

Fine, B. (2002). *The world of consumption.* London: Routledge.

Fine, B. (2007). From sweetness to McDonald's: How do we manufacture (the meaning of) foods? *The Review of Social & Economic Studies, 29*(2), 247–271.

Fine, B. (2013). Consumption matters. *Ephemera, 13*(2), 217–248.

Fine, B. (2020). Framing social reproduction in the age of financialisation. In A. Santos and N. Teles (Eds.), *Financialisation in the European periphery: Work and social reproduction in Portugal* (pp. 257–272). London: Routledge, 257–272.

Fine, B., & Leopold, E. (1993). *The world of consumption*. London: Routledge.

Fine, B., & Milonakis, D. (2009). *From political economy to freakonomics: Method, the social and the historical in the evolution of economic theory*. London: Routledge.

Fine, B., & Saad-Filho, A. (Eds.) (2012). *The Elgar companion to Marxist economics*. Cheltenham, UK: Edward Elgar.

Fine, B., & Saad-Filho, A. (2016). *Marx's Capital* (6th ed.). London: Pluto Press.

Fine, B., & Rustomjee, Z. (1997). *The political economy of South Africa: From Minerals-Energy Complex to industrialisation*. Johannesburg: Wits University Press.

Fridell, G. (2018). Conceptualizing political consumerism as part of the global value chain. In M. Boström, M. Micheletti, & P. Oosterveer (Eds.), *The Oxford handbook of political consumerism* (pp. 309–326). Oxford: Oxford University Press.

Fridell, G., & Walker, C. (2019). Social upgrading as market fantasy: The limits of Global Value Chain integration. *Human Geography, 12*(2), 1–17.

FTEPR. (2014). *Fairtrade, employment and poverty reduction in Ethiopia and Uganda*. Fair Trade, Employment and Poverty Reduction Project, Final Report to DFID. Retrieved from http://ftepr.org/wp-content/uploads/FTEPR-Final-Report-19-May-2014-FINAL.pdf.

Gereffi, G. (2014). Global Value Chains in a post-Washington Consensus world. *Review of International Political Economy, 21*(1), 9–37.

Gereffi, G., & Frederick, S. (2010). The global apparel value chain, trade and the crisis: Challenges and opportunities for developing countries, World Bank, Policy Research Working Paper, no 5281. Retrieved from http://wwwwds.worldbank.org/servlet/WDSContentServer/WDSP/IB/2010/04/27/000158349_20100427111841/Rendered/PDF/WPS5281.pdf

Gereffi, G., Humphrey, J., & Sturgeon, T. (2005). The Governance of Global Value Chains. *Review of International Political Economy, 12*(1), 78–104.

Giddens, A. (1979). *Central problems in social theory*. London: Macmillan.

Gibbon, J. Bair and S. Ponte (Eds.) (2008). "Governing Global Value Chains", special issue of *Economy and Society, 37*(3).

Haiven, M. (2018). *Art after money, money after art: Creative strategies against financialization*. London: Pluto.

Hartmann, E. (2014). The fetish of global competition. *Capital & Class, 38*(184). https://doi.org/10.1177/0309816813514210.

Haug, W. (1986). *Critique of commodity aesthetics: Appearance, sexuality and advertising in capitalist society*. Minnesota: University of Minnesota Press.

ISP. (1995). *Improving manufacturing performance: Report of the Industrial Strategy Project*. Cape Town: UCT Press.

Kallis, G., & Swyngedouw, E. (2018). Do bees produce value? A conversation between an ecological economist and a Marxist geographer. *Capitalism, Nature, Socialism, 29*(3), 36–50.

Meadows, D. (2008). *Thinking in systems: A primer*. London: Earthscan.

Robertson, M. (2020). How and why are things consumed? In E. Van Waeyenberge & K. Deane (Eds.), *Recharting the history of economic thought* (pp. 69–88). London: Routledge.

Southerton, D. (2011). *Encyclopaedia of consumer culture, Volume 1*. London: Sage.

Sturgeon, T. (2000). *How do we define value chains and production networks?* Massachusetts Institute of Technology Background Paper Prepared for the Bellagio Value Chains Workshop September 25–October 1. Retrieved from https://www.ids.ac.uk/ids/global/pdfs/vcdefine.pdf.

Turner, S., Derks, A. & N. Hạnh (2019). Flex crops or flex livelihoods? The story of a volatile commodity chain in upland northern Vietnam. *Journal of Peasant Studies, 46*(2), 276–296.

Warde, A. (2005). Consumption and theories of practice. *Journal of Consumer Culture, 5*(2), 131–153.

CHAPTER 3

Understanding Material Cultures

3.1 Introduction

Consumption is one of the fundamentals of mainstream economics in that it serves as the basis for understanding demand in the core supply and demand framing of the market economy. It also deeply reflects the methodology and methods of the mainstream, not least its commitment to optimizing individuals (utility maximization in particular), and emphasis on the market as a mechanism for grinding out allocative efficiency in equilibrium (or not in case of market imperfections). Empirical evidence, especially as data and econometrics, tends to be used as an external way of testing or applying the theory (does demand decrease with price and at what rate in estimating elasticities, for example). Yet even from the perspective of the marginally self-reflecting consumer, the way in which the mainstream approaches consumption is remarkably shallow if not flawed. Broadly, whilst the mainstream sees consumption as emerging from the aggregated decisions of more or less isolated, 'rational', utility-maximizing individuals, what and how we consume is profoundly affected by social, certainly familial and peer-group interactions, and is also reflected, for example, in national and ethnic differences, for example when it comes to cuisines. Moreover, people tend to be fairly stable in what they consume over periods of time during which prices change, reflecting habits and customs. Rather than exercising demand independently of one another, individuals conform to systematic patternings, which we will designate as (social) *norms* of consumption. These norms differ, or more exactly are

K. Bayliss, B. Fine, *A Guide to the Systems of Provision Approach*,
https://doi.org/10.1007/978-3-030-54143-9_3

defined, across socio-economic categories such as income, race, gender, location, occupation and their intersections and so on, but these norms take form in different ways, in different locations and with different outcomes according to the specific context.

Consumption norms may be reflected in the personal but they are not individual preferences. Rather they emerge from the values, meanings and practices associated with provisioning. They have been described as 'cosmologies shared by others' (Pellandini-Simányi 2014, p. 51) and so they do not differ randomly, and commonalities can be observed across groups of people. These clusterings of consumption patterns across groups of people suggest that the formation of such norms is driven by factors external to, but internalized by, the individual.

The SoP approach considers consumption through the lens of such norms. Indeed, the approach takes identifiable consumption norms as starting point, with the aim of developing a corresponding explanation for how these are reproduced or transformed and seeks to attend to the differentiated meanings and cultures to which such norms are attached. Broadly, for the SoP approach, consumption norms are considered to be derived from the interaction of the elements within and around the SoP itself (agents, relations, structures, processes, cultures in previous chapter) and, with whatever degree of precision, such norms specific to each SoP can be identified. We engage with such norms constantly in daily life, for example, in decisions ranging from what to eat to what to wear and where to live. The SoP approach also seeks to unpack *why* these outcomes emerge and shift over time, why processes are configured in particular forms, and what underlies consumption and production decisions and the relationships between and around them. Examples range from the detailed patterns of housing tenures for different strata, to pointing to increases in over- or under-nutrition across classes and countries (See Chaps. 4 and 5).

How, then, should we examine the forces that shape consumption norms, given that these cannot be reducible to the atomistic decisions of individuals? To understand more about the nature and significance of norms, the SoP approach draws on the notion of material culture (MC) which is derived from a range of academic disciplines, with anthropology to the fore but including cultural studies, sociology, history, geography, philosophy and, at least in principle, political economy. Significantly, anthropology has turned its gaze from the supposedly primitive cultures of the 'new' world, back upon its own 'old' world to understand corresponding cultures, with modern consumption often to the fore—as well as disposal, with garbology a well-established field![1] The essence of MC, as adopted for the purposes of SoP studies, is that it bridges the supposed gap between the materiality of objects and the social (including cultural) relations to which they are attached. For MC, these are distinct but not seen as separate. Rather, cultures and social relations are considered to be created and reproduced by the ways in which people make, design and interact with the objects around them. Graves-Brown (2000), as cited in Bayliss, Fine & Robertson (2017, p. 362), points out that culture is not in our minds nor independently in the external environment 'but rather is an emergent property of the relationship between persons and things'. In other words, an elephant is very different depending on how you see it, and this is even more so in terms of how you engage with it. Important for MC is not just the cultures themselves but how these are created, maintained and evolving. Those who shape, or whatever shapes, cultures are wielding considerable power.

MC can however be difficult to pin down, not least because it can be so deeply ingrained as to be a kind of 'common sense', so much taken for granted that we are not conscious of its significance, at least until it is challenged by external events or internal reflection. The next section explores the nature of MC in more detail. This is followed by a review of MC in connection with social reproduction which shows how neoliberal narratives can perpetuate social divisions. This section draws on a body of work investigating the MC of financialization to show how social inequalities can be perpetuated by specific narratives. The subsequent section sets out a list of criteria for the researcher to bear in mind when seeking to identify MC in practice, a list known as the 10Cs, before the final section concludes. The chapter, then, provides a steer to ways in which the complexities of MC, which are crucial to understanding consumption, might be unpacked and investigated in SoP analyses. Further insights into practical research approaches can be found in the later chapters of this book which look at specific case studies in more detail.

3.2 The Importance of Material Cultures

Goods and services are always imbued (often subtly or unnoticed) with cultural significance. The narrowly defined physical characteristics attached to provisioning, and consumption, are necessarily culturally endowed in the widest and deepest senses, not least because they involve social as well as individual acts. Cultures are not just in people's minds but also engage the relationships within and between peoples and things. For example, ethnographic research into infrastructure highlights the ways in which its provision can have associations that go far beyond the use value of the services provided by the construct itself. According to Appel, Anand & Gupta (2018, p. 3) in their edited book *The Promise of Infrastructure*:

> the roads and water pipes, electricity lines and ports, oil pipelines and sewage systems, are dense social, material, aesthetic and political formations that are critical to both differentiated experiences of everyday lives and to expectations of the future. They have long promised modernity, development, progress and freedom to people all over the world.

In the same way, to repeat the well-known cliché, an elephant can be a more or less exotic/mundane exhibit in a zoo, subject for a TV documentary, a working beast, or an item of worship. The issues, then, are what are cultures, how are they formed and what effects do they have?

On one level, consumption decisions are made according to cultural associations connected to specific objects. This is clearly illustrated by the fashion systems, where consuming is profoundly cultural—what people want and why—and how they understand what and how they get, do not get, or aspire for something more or different. In other sectors, cultural content may be less obvious. In the provision of water, for example, there may seem to be no cultural associations of significance attached to a glass of water from the tap or running a bath. But even where consumption appears to be mundane, why this is so needs to be explained and not taken for granted. Visit a drought or a flood and the cultures attached to water become prominent if not vital. Disputes can rage over the relative merits of private or public provisioning. And, paradoxically, the mundane nature of the domestic supply of water by public authorities can create the opportunity for the more appealing virtues of commercial bottled water, irrespective of the differences in material quality other than being branded. The best-selling brands of bottled water in the USA (after 'private label' own store brands) in 2018 were Coca Cola's Dasani followed by PepsiCo's Aquafina, both of which are, almost certainly unbeknownst even to their most avid consumers, simply purified tap water (Statista 2019). Yet once in the branded bottle the same water has profoundly different cultural associations over and above those putatively associated with quality.

The SoP approach queries not just the nature of such cultural associations but the ways in which these are, or can be, formed. The drivers of cultural associations are diverse. Beliefs, meanings and practices may be heavily influenced by what is in place and how it is experienced, the inertia attached to preferences for what we already know and do, and constructing corresponding justifications and rationalizations. Personal experience is a strong driver of consumption patterns. There is, for example, a strong body of research that shows that adolescent consumption of tobacco and alcohol are linked to peer and parental influence (Kinard and Webster 2010). Similarly, parental behaviour has a strong role in shaping eating patterns (Savage, Fisher & Birch 2007) although humans have the capacity to be distinctly rebellious and imaginative in both their consumption and their cultures, and the relationship between the two.

In addition to personal experience, cultures are generated and perpetuated by varieties of discourses and narratives and are susceptible to direct attempts to exert influence as with advertising and political or other campaigning. Advertising is intended to sell products by endowing them with properties that they may or may not have. What matters is that they are perceived to have them or to have an association with them, such as exotic or value for money, ideally both, or even sinking below the glamorous to the ordinary (it does what it says on the tin) or rising above the constraints of monetary calculation (go on, spoil yourself). The cultures of provisioning go far beyond ethnic, gender, racial and even national characteristics, to embed themselves in the meaning of the goods and services to which we gain access. In the social media age, advertising has moved on so that 'influencers' such as the Kardashians are reportedly paid six figure sums by product promoters for an Instagram post (BBC 2019). Thus, the association with celebrity does not just increase awareness of the product but imbues it with certain glamorous qualities. And, who can doubt, for example, that advertising promotes the notion that different cars are as much gendered as they are indicative of status, nor that the growth of two-car households and the increasing presence of female drivers underpins changes in how cars are presented and understood even if old male stereotypes persist.

However, although all are engaged culturally, just as all have a vote in a democracy, this does not mean that all are equal culturally with regard to how and what cultures are engaged and with what influences and implications. Meanings and experiences (and associated cultures) underpin the activities of those engaged in the operations of SoPs in ways that necessarily offer unequal scope for contested influence over the processes involved and how they generate meanings and are interpreted. There is differentiation in shaping the SoP across agents, for example, across domains such as by producers/consumers, and across gender or race considerations as well as over the relative merits of private versus public as preferred sources of provisioning. It is a matter of different cultures and their content and impact depending upon where you sit along the SoP. Consumers are rarely able to launch an opposition advertising campaign (Three Billboards[2] as the exception that proves the rule). They might have the sanction of boycotting a product or forming consumer associations to pursue what is a culturally determined (and often contested) consumer interest (lower prices and higher quality, or not, potentially versus decent wages and respect for the environment, and local versus global sourcing and so on).

MC is therefore not just a matter for the consumers, but also for those who provide or influence provision, whether in the public or private sectors, global agents, the policymakers, activists, advertisers or the media, and even academia. These also have cultures of their own, alongside those of consumers, clients, citizens or whatever. The choice of language itself (such as 'citizen' as opposed to 'consumer', passive or not over what terrains) indicates different cultural perceptions of provisioning that are far from neutral to our attitudes or understandings. Further, with privatization, for example, the culture of welfare, citizenship and commitment to basic needs and human rights, and more, tends to be displaced by the culture of the consumer, the client, the ability to pay, and a critique of welfare dependency.

MC also extends beyond decisions around consumption and production. A choice of washing powder may be influenced by advertising, by consumption decisions of peers and parents, and the changing world may also shape the way washing powder is provided with, for example, the advance of eco-branding. But the frame of influence of MC goes far deeper, both to reflect and influence the way in which we engage in society. Governments promote cultures to support political initiatives. The narrative of the 'cheating' welfare beneficiary supports a policy of austerity, for example. More than 85% of calls to a 'benefits hotline' to report neighbours of welfare fraud were found to be false accusations (Cowburn 2016), indicative of a substantial disparity between public perception and reality. Similarly a supposed information campaign encouraging the British public to 'Get Ready for Brexit' had the effect of presenting Britain's departure from the EU as an inevitability when this was at the time still subject to major contestation. As the following section demonstrates, MC has been important in normalizing the embedding of financial practices and financial extraction in everyday lives.

For SoPs, consideration of cultures and meanings are not just relevant for the endpoint of consumption itself but are part and parcel of the whole chain of activity and elements outlined in Chap. 2. MC plays a role in shaping the norms, values, meanings and practices associated with provisioning as well as consuming and is significant throughout the different elements of the SoP. Like the product itself and the chain of activity to which it is attached in provisioning, there is a corresponding chain of determinants of MC which do not simply stand to attention at the point of consumption or in the act of buying. In order to understand fully the drivers of consumption, each SoP needs to be addressed by reference to the material and cultural specificities that take account of the whole provisioning system, bringing together production, distribution, access, and the nature and influence of the conditions under which these occur (Bayliss, Fine & Robertson 2013, p. 1).

3.3 Social Reproduction, Financialization and Material Cultures

MC has potentially significant effects on economic and social reproduction, meaning the ways in which society reproduces itself, particularly in how patterns of inequality are passed on across generations (or not). In the 1970s and 1980s, social reproduction debates focused on the gender imbalance within the household and the role of women's unpaid labour in inter-generational transfers. With the increased penetration of neoliberal concepts and practices, debates have shifted to the commodification of aspects of social reproduction. MC has been significant in the normalization of practices associated with neoliberalism. Here the application of MC leads to areas associated with Foucauldian governmentality, as cultures evolve to support specific power structures and practices, even to the point of being self-imposed, and the making of the neoliberal citizen and consumer is finessed.

Neoliberalism has played a strong role in the 'individualizing' and responsibilization of social provision. As Bayliss, Fine & Robertson (2017) point out in relation to financialization, there is growing evidence that financial practices and financial extraction has extended to ever more aspects of everyday life, as is reflected in a burgeoning literature on the subject. Ever-expanding financialization has meant that public services have been transformed into profitable business opportunities. Significant for SoP and for MC, then, are the structures and configurations of economic, political and cultural powers that underpin this normalization. Understanding the material foundations of cultures in effect 'opens the door to comprehending the role of economic and political power shaping [financialized] cultural forms' (Bayliss, Fine & Robertson 2017, p. 362).

Yet, despite the global trends towards greater neoliberalism and financialization, these have been adopted differently across locations and sectors (Fine 2020). The drilling down of global capital to new areas of social reproduction has not been the same and to the same extent always and everywhere. For MC, the question is how has this become normalized in some cases and not others, and in what different ways and why. What are the processes and the narratives that lead to such practices being accepted or rejected, and in what forms and with what effects?

Cultures are shaped in part by policy. International financial institutions (IFIs) have played a strong role in transmitting neoliberal, market-oriented cultures around the world. Ruckert (2010) shows how developments in global aid policy such as the attachment of privatization conditionality to donor funds led to a transformation in social relations and understandings of state functions. Social policies promoted by IFIs have underpinned specific modes of household behaviour. Citizens have been turned into customers with Conditional Cash Transfers combining 'market-oriented provision of social services with subsidies to the poor', serving to 'perpetuate the downloading of responsibility for social reproduction from the state to the private sector and household' (Ruckert 2010, p. 824; Lavinas 2017, 2018). For Ruckert, the goal of social policy is not to offer protection from the market but to integrate the poor into market (and financial) structures as has also been the goal of the financial inclusion initiatives following on from the now discredited Grameen bank 'revolution' for the poor (Bateman, Blankenburg & Kozul-Wright 2018).

Indeed, in order for households to engage fully in the financialized world, governments have adopted programmes to promote financial inclusion and improve financial literacy. More financially literate consumers are expected to impose competitive disciplining effects on providers of financial and other services. However, such training also aims to instill a culture of self-responsibility for financial management. By extension, this generates a culture of individualized responsibility at the expense of more collective means of provision (Santos 2017). Lavinas (2018) shows how financial inclusion has been turned around to generate indebtedness for poor households and profits for lenders on that debt.[3]

Pervasive in the financial inclusion narrative is the transfer of risk to the individual from the collective, with the financial sector supplying to individuals and households an ever wider range of complex financial products and services. Santos (2017) uses the example of pensions to show that workers need to make their own arrangements for old age to replace collective state provision. Individual risk-taking is encouraged under the auspices of financial inclusion. Such cultures can have the effect of further marginalizing the vulnerable by attributing their condition to bad decisions and rationally, even morally, deviant behaviours, and these are typically gendered effects.

In short, financialization in general has profoundly influenced the practices and effects of social reproduction and, as a result, the corresponding MCs attached to consumption. Such shifting cultures of social reproduction are also evident in the housing sector. In the UK as in much of Europe, the preferred form of housing tenure has long if unevenly (and increasingly) been owner-occupation, in conformity with the squeeze on support for social housing and the rise of more readily available mortgage finance. This preference is not a delusion in the sense that there have been material benefits to homeowners from capital gains in homeownership as well as a concerted attempt to promote this as the preferred tenure through government and media campaigns. But the result has been, if again unevenly across locations, an unprecedented rise in house prices and greater inequality in access to affordable housing, a failure to deal with inadequate housing supply, and an increase in private renting against historical trends, and with owner-occupation out of the reach of many. And housing benefit for private renting across Europe has increased substantially at the expense of expenditure on the provision of social housing, essentially using the hard to house as an intermediary in the transfer of state subsidies to private landlords, whilst reinforcing the notion of supported private tenants as scroungers (National Housing Federation 2017)

MC sheds light on changing meanings and values and their social effects. Each of the agents in the SoP, from producers to consumers and the state will bring (and take) specific understandings and meanings to (from) the processes of provisioning, and these may be contradictory or even involve direct conflict over outcomes and meanings/cultures. One person's flexible (self-employed) work force is another's labour exploitation or one person's home can be another's financial asset. The way in which these themes intersect and shape consumption will depend on the specific context of provisioning and its social grounding.

3.4 Framing Material Culture—The 10Cs

SoPs need to be specified through a combination of analyses of the processes of provision outlined in the previous chapters and their mutually constitutive interactions with MCs. There is an intimate, integral two-way relationship between material provisioning and material cultures (and varieties of discourses) attached to SoPs, with each conditioning the other.

Framing and specifying MC presents serious challenges. As mentioned, cultures are often so ingrained it can be difficult to see them as anything other than ordinary and taken for granted. Many aspects need to be understood, taking account of cultural specificity, the efficacy of cultural properties, agent reflexivity, and the co-constitution of subject and object, while 'continu[ing] to emphasize the materiality of social relations and the constraints involved in processes that also operate "behind the backs" of the relevant agents' (Bayliss, Fine & Robertson 2017, p. 362, citing Jessop and Sum 2001, p. 94). Koch and James (2020) demonstrate the subtleties and complexities of cultures in their study of outcomes under neoliberal welfare provision under austerity. They highlight the position of welfare workers whose role as advisers to 'clients' is torn between competing objectives with both commitments to social justice alongside 'logics of deservingness and the poor' (p. 12) within the state apparatus. Indeed, 'the "state of the welfare state" produces multiple, shifting and contested boundaries of inclusion and exclusion that require careful ethnographic attention' (p. 12).

To facilitate the deconstruction of the extensive influences on agents' subjectivities in practice, the factors that shape cultural systems have been grouped by Fine (2013) under ten headings (known as the 10Cs), and these interact with each other in complex and diverse ways. The aim of the 10Cs is to provide pointers to capture or to bridge both the complex nature of material cultures (the natures and contents of meanings) and the ways in which they are forged through, or conditioned by, the material and social processes with which they are embroiled.[4] The purpose of the 10Cs is to provide guidance for the full comprehension of consumption cultures and how these are constituted through the identification of ten characteristics common to all such cultures.

So MC is a feature of SoPs, for example, the supplying of water itself could be perceived as a public service or as a profitable industry. Similarly, depending on context, there may be different perceptions of publicly provided, as opposed to bottled, water. But also important for understanding the structures that underpin society is to determine from where these cultures themselves come. How is MC made? Who and what determines how we perceive our water beyond our own reflections? For also the rationale for the 10Cs is to provide space to consider agents' capacities for reflection and resistance and so avoid any tendency to exaggerate the role of cultural hegemony as self-supporting and reproducing, let alone imposed by a more or less hidden thought or cultural police. But, as it were, even if we make our own beliefs and cultures, we do not do so in circumstances created by ourselves.

Consider, though, in relation to financialization, that resistance has been remarkably muted, at least in the UK (Bayliss, Fine & Robertson 2017). Analysis of MC can shed light on why this is the case. Fine (2017, p. 372), for example, uses the 10Cs to show the 'diverse pathways by which the Foucauldian governance of the financialization of everyday life prevails'. So the idea is to develop an understanding not just of the ways in which specific understandings become embedded in everyday lives but also the structures and configurations of economic, political and cultural powers that underpin both such normalizations and the potential for resistances and alternatives to them.

The 10Cs have become part and parcel of understanding material cultures in applying the SoP approach. While there is a strong resonance to its origins within Marxist political economy, the use of the 10Cs framing is compatible with different approaches and methodologies (in common with the SoP approach more broadly) and is applicable to a range of subject matter (not just consumption). The 10Cs relate to a series of attributes of material cultures which will apply to varying degrees or, more exactly, in varying ways, depending on the circumstances. Not every 'C' will be equally relevant in all cases but the aim is to provide a useful checklist for interrogating material cultures of specific provisionings, as set out below.

Constructed—the cultural systems attached to consumption are constructed in that they are influenced by the material practices of the SoP and by agents in the SoP. We have met many examples of this already beginning with the discussion of commodity fetishism in Chap. 2, whereby a culture for understanding commodities arises out of market relations that conceal those attached to production. Such private provisioning for profit also tends to be attached to the notion of consumption on the basis of the capacity to pay as opposed to public provisioning, outside of the market, whereby the presumption is one more of human rights, capacity for citizenship, merit goods, basic needs or whatever. So how provisioning is constructed profoundly affects the way it is perceived. Another example is the notion of produce as 'home-made' or, if commercially provided, the claim to be as good as home-made even if not. Similarly, there can be claims of technical wizardry in goods provided beyond the capabilities of the home by virtue of being commercially manufactured. But, as revealed by commodity fetishism, and the analysis in Chap. 2 that takes Haug's aesthetic illusion as critical point of departure, what enters from the construction of provisioning into culture is not a direct, let alone a full nor necessarily accurate reflection of the provisioning itself. Constructed means that these cultures are shaped by the material factors of the SoP and by the agents within it. In our 10Cs, we consider cultures to be Constructed by the agents of the SoP with meanings that may be Construed differently by households and a whole variety of other influences however much engaged directly with provisioning and its culture.

Construed—notwithstanding the way the SoP is constructed, consumers (and other agencies along the SoP) necessarily have to make sense themselves of the ways in which they engage in and experience provisioning. This is obvious in the case of advertising, for example, in which belief can hardly be suspended given some of the claims made (although it is not the veracity of claims that ads necessarily want as opposed to image-inducing purchase). Consumers, and others, bring their own knowledges and beliefs, experiences and practices to provisioning. In short, construal relates to the ways in which meanings are internalized and reflected upon by consumers, households and other SoP agents. These can float free from the material properties of the objects themselves. The process of construal is influenced by a multiplicity of factors and these are derived from context. Sources of experience and knowledge are reacted to, or against, and imbued with meaning rather than simply received passively by the consumer. So experience may lead to perceived associated cultural properties as attached to some forms of provision. For example, advertisers promote diverse images of masculinity, both promoting and challenging stereotypes. These messages will be interpreted and acted upon differently across consumers. Similarly, the right-to-buy housing policy in the UK, which gave some tenants of social housing the legal right to buy their home at a discount, can lead to a change in culture with shelter then increasingly construed as a financial asset at the expense of, or alongside, its being a home. Thus, the house which may be physically unchanged, becomes private property as opposed to a residence in the form of collectively provided and available social housing.

Commodified—to greater or lesser degrees, cultures may be influenced by commodification (as the way in which culture is constructed) even if the good itself is not. Even where consumption is not commercialized its culture may be constructed in relation to its opposition to commerce with this as a selling point (eg home-made) or as a mode of management. The process of commodification serves to frame ways of thinking about and interpreting what is consumed. The logic of pushing the boundaries of commodification is associated with neoliberalism and financialization, for example, with health services in England framed as discrete packages of provision, each with a separate price tag if not bought and sold (see Chap. 4). This is present even in publicly provided health services as a presumed means to promote cost efficiency. Commodification promotes a shift in thinking around the way in which a good or service is produced and/or consumed and conceived. The commodification of water is a useful example. In England and Wales, volumetric metering where billing relates to consumption has replaced a charging system where bills were previously related to property value as a loose approximation for wealth. Such a transition is intended to change the way in which water is perceived by instilling a sense of scarcity. Preservation and reductions in usage are promoted using financial mechanisms, known as 'demand management' as households are induced to have a financial incentive to lower water consumption.

Conforming—precisely because culture is subject both to commodification and construal in the circumstances dictated by SoP provisioning, so corresponding cultures tend to conform to the dictates of the common senses attached to the experiences of consumption as opposed to engaging with them critically (although this is also possible, see below). Certain narratives tend to dominate, for example, owner occupation is seen as obviously better than social housing, competition generates efficiency and so on, especially given that the public sector is deprived of resources and is responsible for the residualized and hard to serve. Thus, cultures conform in the sense of relying upon a kind of common sense but this raises the issue of *why* common sense has the content that it does, how does it change and what are the structures, relations agencies and processes by which it is reproduced or transformed.

Contextual—cultures of consumption, like SoPs themselves, differ across time and place so that what is consumed is not only located in specific circumstances (high or low price, good or bad quality) but these are associated with particular and variable meanings to the consumer (e.g., an item of clothing may have different significance depending on who is wearing what, when and in what situation). That corresponding cultures are contextual is acknowledged and addressed by ethnographic studies, and reflected, for example, in different eating habits and preferences, especially around religious proscriptions of what is or is not fit to eat. Further, one person's necessity may be another's luxury and the distinction may change over time, location and across income levels. The 'same' object of consumption, such as a McDonald's hamburger, can have both different material and cultural content in different situations and these be differently determined too (Fine 2007).

Contradictory—different agents and forces compete, or complement and add complexity, to give content to the cultural systems, and these may provide a stimulus in different directions. Cultures rely on sometimes opposing social forces and tendencies. For example, we may be pressured to adopt more healthy lifestyles while at the same time surrounded by triggers that promote an unhealthy diet. Similarly, we are simultaneously encouraged both to spend and to save. Fine (2017) provides a further example, citing Dodd (2016, p. 183):

> Whilst neoliberal financialization may, ultimately, be characterized by the pathological and limitless collapsing of all social values into and under the ruthless quantitative measure of economic value, it is also, contradictorily, dependent on the forms of innovation, social reproduction and subjectivity generated in the intervals.

Thus, while neoliberalism is founded to some degree, if far from exclusively, on the narrowing of social interactions into a common economic measure, the monetary, at the same time its expansion relies on imaginative new applications. If everything is reduced to money, and monetary calculation, so this opens up the culture of rejecting this tendency, as most obviously expressed in the cliché that money cannot buy love, and something most obviously brought explicitly into the open in the artistic world (artist's integrity against commercial imperatives to survive or prosper), with Banksy's self-shredding artefact upon purchase.[5] From Jesus' assault on the moneylenders to the anomie directed at the 1%, there are long-standing traditions of both acknowledging, even conforming, to the monetary and also rejecting it. The issue is more where the shifting and complex lines between the two are drawn, as with the deploring conformity to the commercialization (and promotion and Americanization) of festivals from Christmas to Halloween, and how and with what effects on MC.[6]

Chaotic—material cultures draw together (or not) a multiplicity of practices and influences across a multiplicity of dimensions which are reflected on by households going about their daily life and so will be riddled with inconsistencies as choices and activities are engaged in practice, with reasons, feelings and motivations involving shifting, even innovative, combinations across them. This does not mean that there is no rationale for corresponding outcomes, but that these will be contingent on how practices and cultures mutually condition one another. Ex ante predictions are difficult to make.

Closed—there are unequal and differentiated roles in the influences exerted over cultures of consumption. Who gets to shape cultures is not necessarily equally open to all. This may be down to access to information but also is shaped by power relations. The shaping of cultures tends to be disproportionately exercised by a few, not least through the media. On one level, there are advertisers and 'influencers', particularly more recently on social media, but significant pressure also comes from the state through regulation and legislation. Corporate power is manifest in many ways from political donations to revolving doors between industry and government. In the UK the Big 4 consulting firms advise the government on policy across many sectors and advising on post-crisis financial regulation has more than made up for the minor setback of the financial crisis in protecting financial interests (Brooks 2018).

Contested—Chap. 2 set out the core elements of SoP analysis and the theoretical background. It was stated that the agents in the provisioning system likely have interests that are incompatible (such as between capitalist and worker). Similarly, agents will have an interest in projecting conflicting cultures, as orthodoxy comes into contact with resistance pushing for radical rethinks of cultural framings. Significant here are major protests such as the Occupy movement and global activity around climate change. Contestation may be both indirect and in terms of the conditions attached to the material practices along the chain and the way these are endowed with meaning. What is meant by Fair Trade, for example? In the context of growing dissent around the ethics of corporate practices, major companies are seeking to portray a narrative of social inclusion and environmental sustainability, for example, with major fossil fuel companies professing their commitment to renewable energy.

Collective—MC operates on a group rather than, or just at, individual levels. As observed above, cultures tend to operate across clusters and communities. Moreover, rarely does an individual significantly shape MC. Usually the voices that shape both orthodoxy and contestation are collective, more or less deliberately so, operating through the wheels of politics and lobbying or in group protest movements.

In devising the 10Cs, many other Cs have been suggested. One in particular has been Class. But this should not be privileged as an element in cultural systems of consumption any more than gender, ethnicity or whatever even though these may be vital determinants. As consumption is driven by the commodity form, it is the capacity and willingness to pay that is the most important proximate quantitative determinant. Whilst income and lifestyle do have some correspondence to class position, this is sufficiently loose and variable across SoPs that class as such is not necessarily an immediate determinate, or defining characteristic, of corresponding cultural systems. Class, then, is an important determinant of consumption and is part of the SoP (see above) but it is not inevitably a proximate determinant of corresponding cultural systems, and its presence can be filtered through the other 10Cs, possibly overtly through the Collective, the Construed, and the Contested for example. Think TVs, washing machines, haircuts, mobile phones and so on. It would be hard to find a class structure in these that prevails, like writing in a stick of rock, especially distinctive from income levels and other determinants (Fine 2013).

3.5 Conclusion

Consumption does not arise spontaneously nor randomly but rather through patterns that can be observed across different groups in society in ways which are specific to the context and to the commodity, or good more generally, in question. Hence analysis of consumption needs to be oriented around norms and inevitably requires some framework for understanding how these arise, and change. MC offers considerable insights into the drivers of consumption norms and the roles of the agents in the SoP in their development. MC is important in determining the content of the apparent 'common sense' around provisioning (e.g., private is better than public, or vice versa) as well as when and how it might be contested or changed.

For SoP analysis, each good and service will be tied to an integral material system of provision as well as to wider cultural influences but in ways that are not rigidly predetermined nor necessarily immediately obvious. Such cultural content is also subject to wider considerations that range far beyond the immediate provision of the good itself (such as gender, class and national identity). MC attached to the provision of goods and services not only shapes the quantities and ways in which they are consumed but also how they are understood in their passage from production to consumption.

Operationalizing such analysis is not straightforward and the 10Cs offer a systematic collection of themes to be considered. The list includes aspects which emerge from the materiality of the SoP itself and of the activities of agents in provisioning (Constructed and Contradictory). Other themes relate more closely to the reflexive properties of the cultures themselves. These are associated more with the way in which the individual engages with the SoP (Construed, Chaotic and Conforming). The remaining Cs bridge the gap between materiality of the SoP and reflexivity (Commodified, Contested, Collective, Closed and Contextual).

Use of the 10Cs, whilst originating from extension of the SoP approach, is compatible with applications to a range of subject matter other than consumption. An understanding of the way in which cultures emerge and how they are rooted in narratives in discourse sheds light on the role of political and economic power in shaping cultural forms and the way these become legitimized. MC is important for understanding significant shifts within neoliberalism and capitalism more broadly. These are not reducible to a collection of policies but are more deeply embedded in cultures, as, for example, financial calculation is normalized in ever more areas of everyday life (Fine and Saad-Filho 2017). As a result, the politics of MC is both vital and complex, and needs to be unpicked if progressive change is to be brought about, with a need to challenge what is provided, how and with what understandings. Such is particularly apposite as the current phase of neoliberalism seeks to extend social provision—with rights to health, education, even income—on the basis of expanding, often state-supported, private provision. Whilst, then, there has been a swing towards mainstream acceptance of the goal of Universal Health Care (UHC)—in that sense a battle has been won—there still remains major issues and differences over the war of what is understood by UHC and how it is to be delivered and by whom (see McKee, Balabanova, Basu, Ricciardi & Stuckler (2013) and Stuckler, Feigl, Basu & McKee (2010)). As emphasized throughout this chapter, and before, there are no universal answers given the specificities of provisioning associated with individual SoPs, a point reinforced by consideration of the interactions between SoPs and their MCs. As a result, case studies follow in Chaps. 4 and 5, to illustrate how the SoP approach can be operationalized as opposed to picking a framework off the shelf into which SoPs and MCs can be neatly fitted.

Notes

1. This was neatly remarked by Witold Rybczynski (July 5, 1992) in his article 'We Are What We Throw Away' (https://www.nytimes.com/1992/07/05/books/we-are-what-we-throw-away.html) which piggy-backs on (one-time Mayor of Belley) Brillat-Savarin's often misquoted dictum, derived from Feuerbach as 'You are what you eat'. It should in fact more accurately and appropriately read, 'Tell me what you eat and I will tell you what you are', which works to some degree for your rubbish bin too.
2. See https://www.youtube.com/watch?v=Jit3YhGx5pU
3. And on the triggering role of conditional cash transfers, Lavinas (2013).
4. The 10Cs approach has been adopted as a general approach to material culture as is apparent from its application to public provision and social policy (Fine 2002, 2014), and to topics such as identity (Fine 2009a, b), the ethics of economics (Fine 2013), and (international) legal expertise (Fine 2016).
5. See https://www.google.com/search?q=banksy+self+shredding+art&oq=banksy+self&aqs=chrome.4.0j69i57j0l6.6953j0j7&sourceid=chrome&ie=UTF-8
6. Note that the commercialization of Christmas was covered in Fine and Leopold (1993), not least the breach of the US tradition of not starting festive shopping until after Thanksgiving (the last Thursday in November which ultimately proved too late for commerce to bear once falling upon the 30th).

References

Appel, H., Anand, N., & Gupta, A. (2018). Introduction: Temporality, politics and the promise of infrastructure. In N. Anand, A. Gupta, & H. Appel (Eds.), *The promise of infrastructure* (pp. 1–38). Durham: Duke University Press.

Bateman, M., Blankenburg, S., & Kozul-Wright, R. (2018). *The rise and fall of global microcredit: Development, debt and disillusion*. London: Routledge.

Bayliss, K., Fine, B., & Robertson, M. (2013). *From financialisation to consumption: The systems of provision approach applied to housing and water*. FESSUD Working Paper Series, No. 2. Retrieved from http://fessud.eu/wp-content/uploads/2013/04/FESSUD-Working-Paper-021.pdf.

Bayliss, K., Fine, B., & Robertson, M. (2017). Introduction to special issue on the material cultures of financialisation. *New Political Economy, 22*(4), 355–370.

BBC. (2019). *How much does Kylie Jenner earn on Instagram?* BBC Retreived from https://www.bbc.co.uk/newsround/49124484

Brooks, R. (2018). The financial scandal no one is talking about. *The Guardian Long Read*. Retrieved from https://www.theguardian.com/news/2018/may/29/the-financial-scandal-no-one-is-talking-about-big-four-accountancy-firms.

Cowburn, A. (2016). More than 85% of public tips on benefit 'frauds' are false. *The Observer Newspaper*. Retrieved from https://www.theguardian.com/society/2016/feb/27/false-benefit-fraud-allegations.

Fine, B. (2002). *The world of consumption*. London: Routledge.

Fine, B. (2007). From sweetness to McDonald's: How do we manufacture (the meaning of) foods? *The Review of Social & Economic Studies, 29*(2), 247–271.

Fine, B. (2009a). The economics of identity and the identity of economics? *Cambridge Journal of Economics, 33*(2), 175–191.

Fine, B. (2009b). *Political economy for the rainbow nation: Dividing the spectrum?* Paper prepared for 'Making Sense of Borders: Identity, Citizenship and Power in South Africa', South African Sociological Association, Annual Conference, June/July, Johannesburg. Retrieved from http://eprints.soas.ac.uk/7972/1/sasa_benfine.pdf.

Fine, B. (2013). Consumption matters. *Ephemera, 13*(2), 217–248.

Fine, B. (2014). *The continuing enigmas of social policy*. United Nations Research Institute for Social Development Working Paper Series, No. 2014-10. Retrieved from http://www.unrisd.org/unrisd/website/document.nsf/(httpPublications)/30B153EE73F52ABFC1257D0200420A61?OpenDocument.

Fine, B. (2016). The systemic failings in framing neo-liberal social policy. In T. Subaset (Ed.), *The great financial meltdown: Systemic, conjunctural or policy created?* (pp. 159–177). Cheltenham: Edward Elgar.

Fine, B. (2017). The material and culture of financialization. *New Political Economy, 22*(4), 371–382.

Fine, B. (2020). Framing social reproduction in the age of financialisation. In A. Santos & N. Teles (Eds.), *Financialisation in the European periphery: Work and social reproduction in Portugal* (pp. 257–272). London: Routledge.

Fine, B., & Leopold, E. (1993). *The world of consumption*. London: Routledge.

Fine, B., & Saad-Filho, A. (2017). Thirteen things you need to know about neoliberalism. *Critical Sociology, 43*(4–5), 685–706.

Kinard, B., & Webster, C. (2010). The effects of advertising, social influences, and self-efficacy on adolescent tobacco use and alcohol consumption. *The Journal of Consumer Affairs, 44*(1), 24–43.

Koch, I., & James, D. (2020). The state of the welfare state: Advice, governance and care in settings of austerity. *Ethnos*. https://doi.org/10.1080/00141844.2019.1688371.

Lavinas, L. (2013). 21st century welfare. *New Left Review, 84*, 5–40.

Lavinas, L. (2017). *The takeover of social policy by financialization: The Brazilian paradox*. London: Palgrave Macmillan.

Lavinas, L. (2018). The collateralization of social policy under financialized capitalism. *Development and Change, 49*(2), 502–517.

McKee, M., Balabanova, D., Basu, S., Ricciardi, W., & Stuckler, D. (2013). Universal health coverage: A quest for all countries but under threat in some. *Value in Health, 16*, 539–549.

National Housing Federation. (2017) Public expenditure on housing: The shift from capital spend to housing allowances. A European trend? Research briefing. Retrieved from http://s3-eu-west-1.amazonaws.com/pub.housing.org.uk/public_spending_housing_europe_uk_briefing.pdf.

Pellandini-Simányi, L. (2014). *Consumption norms and everyday ethics*. London: Palgrave Macmillan.

Ruckert, A. (2010). The forgotten dimension of social reproduction: The World Bank and the poverty reduction strategy paradigm. *Review of International Political Economy, 17*(5), 816–839.

Rybczynski, W. (1992, July 5). We are what we throw away. *New York Times*. Retrieved from https://www.nytimes.com/1992/07/05/books/we-are-what-we-throw-away.html.

Santos, A. (2017). Cultivating the self-reliant and responsible individual: The material culture of financial literacy. *New Political Economy, 22*(4), 410–422. https://doi.org/10.1080/13563467.2017.1259302.

Savage, J. S., Fisher, J. O., & Birch, L. L. (2007). Parental influence on eating behavior: Conception to adolescence. *The Journal of Law, Medicine & Ethics: A Journal of the American Society of Law, Medicine & Ethics, 35*(1), 22–34.

Statista. (2019). Sales of the leading bottled still water brands in the United States in 2018 (in million U.S. dollars). *Statista portal*. Retrieved from https://www.statista.com/statistics/188312/top-bottled-still-water-brands-in-the-united-states/.

Stuckler, D., Feigl, A., Basu, S., & McKee, M. (2010). The political economy of universal health coverage. Background paper for the Global Symposium on Health Systems Research, November 16–19. Montreux: Switzerland.

CHAPTER 4

Insights from Operationalizing the Systems of Provision Approach

4.1 Introduction

This chapter draws on well-established applications of the SoP approach to give a flavour of the way it has been applied by researchers as well as the issues addressed and the nature of the main findings. The purpose of this book has been to present the SoP approach in a simple and persuasive fashion—how and why it originated, what are its core elements, how it draws upon but is distinct from, and even critical of, other approaches to consumption and provisioning, how it has evolved, and how others might operationalize the approach in their own understanding and work. An important part of this exercise is to show how the approach has been used in practice. In advance, especially given our own use of the SoP approach across a number of diverse case studies upon which to draw—and thirty years of contributions from others as well—we were confident that a chapter covering a select review of SoP case studies as a guide on how to operationalize the SoP approach would prove easy to draft, almost writing itself. In the event, it proved the most challenging of the chapters, leading us to consider advising 'just get on your SoP bike and pedal'!

It is worth rehearsing why this situation arose. The difficulty, as exemplified by the preceding chapters, does not derive in principle from identifying, and potentially putting together, those core elements that comprise a SoP. These include: the structures, agents, processes and relations that constitute the chain of activity linking production to consumption; the consumption norms that they generate; the material cultures to which

K. Bayliss, B. Fine, *A Guide to the Systems of Provision Approach*,
https://doi.org/10.1007/978-3-030-54143-9_4

they are attached; and situating the analysis in context of both the grander factors involved (globalization, neoliberalization, financialization, etc) and the specifics of the SoP itself as socially, historically and locationally determined. Where the difficulty does occur is where to begin in deploying these elements in an overview of SoP applied research. Is it with production or consumption, with culture or norms of provision, is it with the agents or the processes, and so on?

This conundrum, in turn, reflects a more general issue—the distinction between the order of investigation (by SoP researchers) and the order of presentation (to others). In retrospect, it is easy to realize in our own work, and presume this is true of others, that before we present a SoP we have spent considerable time investigating what it is, why and how, and then come to a view on how best to explain this to others, and this is liable to be in a way that does not coincide with how the investigation occurred. This is almost inevitable given that presentation is linear—one thing after another—whereas the investigation goes from one factor to another, and back again, developing and refining in understanding how different factors interact with one another. Such is necessarily the case given the SoP approach is systemic—taking a view of the whole in light of the (socially grounded) integral interaction of its constituent parts (which can themselves be cut in different ways, not least by the core elements of the SoP itself).

To a large extent, this issue of investigation/presentation dovetails with, but is distinct from, identifying the boundaries of the SoP itself. Here, we have argued that this is primarily driven by the research question being asked as well as by the workings of SoPs in practice themselves. Necessarily, research questions, and case studies which use the SoP approach, are highly diverse. But there are some common starting points that we can identify with corresponding case study illustrations.

In the literature there is a range of engagement with the SoP approach, from dedicated studies that explicitly develop a more or less full and deliberate SoP analysis. At the other extreme are those contributions that draw on the SoP approach, or related framings and which can be readily slotted into a SoP framing. This chapter, then, is far from an exhaustive review of the SoP literature, but draws on select applications to highlight some of the core themes that have arisen to date. Our concluding chapter takes up more, and generally more recent, studies to locate the SoP approach within the evolving field of consumer studies.

As suggested already, the challenge here is to decide what goes where and where to start. In retrospect, it is worth mentioning that, going into the 1990s, the range of SoP applications expanded from private commodity consumption to public consumption—especially as the neoliberal drive to denationalize was peaking but being augmented by an accelerating pressure for privatization through new forms of state/private partnerships.[1] In order to capture the distinction between public and private consumption, the clumsy term public sector system of provision was introduced and its corresponding, equally clumsy acronym, PSSoP—see Chap. 1 for an account of how this fits within the overall development of the SoP approach.

First though, the following section explores some of the contributions of the SoP approach to wider areas of scholarship, including in consumption studies and understandings of social policy. Section 4.3 turns to explore more specific SoP applications. Our focus here is on selected areas of everyday life covering housing, water, health services and fashion. Another key aspect of the everyday is food which is the subject of the following chapter. The concluding section remarks, in light of the case studies, on how the SoP approach unveils what is hidden from the consumer at the point of consumption.

4.2 Addressing the Horizontal and Scholarship

The SoP approach is not only prompted by empirical and policy issues around understanding who gets what, and how but also why as well as how this may be changed. As fully laid out in earlier chapters, the SoP approach's own intellectual origins derive from critically drawing upon and integrating contributions from across the social sciences to allow for genuine interdisciplinarity in addressing consumption and provisioning more generally. Accordingly, the SoP approach is able to situate critically the contributions of different social sciences to the study of consumption. It observes the tendency to overgeneralise from specific factors or case studies, without properly differentiating the material and cultural distinctions that are attached to one SoP as opposed to another.

Especially to the fore, given political economy as its starting point, has been the SoP approach's exposure of the excruciating weaknesses of mainstream economics with its core starting point of given individuals (with limited identity, let alone inventiveness, other than to maximize utility) and given goods (with equally limited confinement of their substance to some unspecified physical capacity to serve utility). Under 'economics imperialism',[2] the colonization of other social sciences by economics, everything from familial relations through drug addiction and crime to the formation of institutions has been one-dimensioned into this narrowly confined supposed economic rationality.[3] As a result, the SoP approach has been particularly adept at bringing in other factors to bear (critically drawing upon other social sciences) as determinants of consumption broadly conceived. SoP research has typically been oriented around vertically integrated case studies within sectors in materially and culturally defining and determining the factors which shape consumption, thereby providing critical counter-examples to the mainstream's 'horizontal'[4] explanations, such as dependence on utility maximization.[5] And, by the same token, the SoP approach, from its outset, took postmodernist-style interpretations of consumption as critical point of departure, seeking to ground the inventively subjective consumer in differentiated material conditions and cultural contexts of provisioning—see, for example, discussion of advertising and its material culture (store wars) in Chap. 1.

The SoP approach has also engaged with more general scholarly issues involving consumption. For example, one of the earliest contributions from within the SoP approach concerned a critical assessment of the notion of consumer society or, equivalently, consumer revolution. Such terms have been widely used both historically and in regard to contemporary capitalism both to identify consumption as a, if not the, driving force for economy and culture and either to praise it (as engine of growth) or to condemn it (as one-dimensioning ethos and cultures to personal gratification). Thus, for example, within the history of the industrial revolution literature, Fine and Leopold (1990) provided a critique of McKendrick's highly influential study of the UK's industrial as consumer revolution. McKendrick drew primarily upon the role of Wedgwood pottery, and the inventiveness and marketing skills of Wedgwood himself to draw attention away from the industrial revolution as production and to praise the entrepreneur and the market as levers of development and satisfaction. Quite apart from pottery as a special and limited case, serving an elite with limited trickle-down at the time, the rise of tea drinking might be thought to have at least as much to do with the British Empire and tea plantations (and the sweetness of sugar slavery too) in creating consumption habits as opposed to the entrepreneurial spirit in making and serving consumers.[6]

The point here is both to follow consumption back to its origins in production and to contextualize, and not generalize, from what are liable to be differentiated SoPs in content and impact. This is demonstrated in more detail in the cases that follow, below.

Significantly, across the historical (and contemporary) literature, it is possible to find the notion of consumer society used, whether to be welcomed (growth-inducing) or deplored (false needs), as applicable to most periods and places where there has been growth of, or spurts in, consumption. The negative literature on consumer society or consumerism[7] has tended to focus on the distinction between false (in terms of their artificial creation, e.g., through advertising) needs and true needs. From the SoP perspective, and other critical approaches, the distinction between false and true needs is crude even if not necessarily to be rejected if contextually grounded. There are no readily defined universal needs beyond basic survival other than in the most abstract forms.[8] So what constitutes basic human needs is inevitably socially constructed (e.g., as in theories of relative as opposed to absolute poverty). More important than interrogating the dualism between false and true needs is to deconstruct how those needs are part and parcel of much more complex material cultures of consumption as targeted by the 10Cs approach, see Chap. 3.[9]

In this vein, the SoP approach has made critical contributions across the evolving work of Amartya Sen (on famine, entitlements, capabilities and so on) and ethics more generally,[10] drawing upon the approach's own understanding of how the meanings of needs are socially constructed (if not necessarily delivered as fully as is implied to be necessary). Such needs, from the SoP perspective, are sector specific in social and historical context. One of Sen's major contributions was to link famine to generalized failures of entitlement to food. He suggests that democracy and free press might offer some guard against entitlement-failure induced famine. But it is far from clear that this would be a remedy for other entitlement failures or even for the food-related diseases of affluence for which there is excessive, not deficient, consumption, see Chap. 5. More generally, the understanding and social construction of capability provisionings (consumption) and meanings of these and other basic needs and/or human rights are open to be addressed by the SoP approach in a deliberately sector-specific fashion.[11]

This is to engage, at least indirectly and implicitly, with the issue of social policy. In this, the SoP approach was used as the perspective from which to launch a scathing attack on the leading approach to the field, the Esping-Andersen's Welfare Regime Approach, WRA.[12] The WRA initially proposed three regimes of welfare provision corresponding to liberal (e.g., the USA and UK), the Bismarckian (authoritarian paternalism, Germany), and the social democratic (the gold standard of Scandinavia). Subsequently, many other regimes have been proposed in response to ill-fitting case studies within countries in explaining different features of different sectors of social policy or in response to ill-fitting case studies from other countries, especially the Mediterranean, Latin American and East Asian, as these failed to fit the original. Whatever Esping-Andersen's original intentions, to the fore, from the SoP perspective, was the literature's illegitimate strait-jacketing of social policy both within and between countries into ideal-type, empirically identified, off-the-shelf welfare regimes. In addition, with whatever validity as it emerged, Esping-Andersen's approach had been very much the creation of the post-war boom period, seeking to distinguish different systems of welfare provision in (Keynesian-style) contexts that were already being discarded in the 1990s when the WRA was first put forward.

Indeed, the WRA has been very backward looking, seeking to understand the neoliberalisation of social policy through welfare regimes themselves in the context of these not being, as it were, Keynesian any more with the general, and commendable, goal of restoring those policies of the past if not upgrading them to the social democratic as opposed to cutting and commericializing in deference to austerity. This is, however, to fail to treat neoliberalism on its own terms, and especially the direct and indirect influences of financialization upon social policy (which has simply been disregarded in the WRA literature with at most an acknowledgement of how policy responds to austerity).

In contrast to the WRA, for the SoP approach, social policy is anchored in empirically grounded study of differentiated analyses by sector and by country and, most recently, within the context of globalized, financialized and neoliberalized provision, as demonstrated in the case studies outlined below. The approach draws on (and contributes to) scholarship which is concerned with 'horizontal' framings of social outcomes such as the issue of 'social reproduction' relating to the ways in which social inequalities are transmitted from one generation to the next. This concept has mainly been deployed by Marxist political economy to distinguish it from *economic* reproduction—the production and circulation of (surplus) value in a capitalist economy, itself embedded within social reproduction.[13] Particular and original emphasis has been placed upon how the workforce is reproduced, beyond the provision of the wage and waged goods, not least through domestic (i.e. household labour). Social reproduction, though, ranges across much more than the wage-provided goods and domestic (household) labour. It also encompasses the role of other classes (who are also reproduced after all) and of the state, not only as provider, but in broader economic, political and ideological relations. Much of this can be broached through consideration of those SoPs for (private and public) consumption that are in major part constituent of social reproduction—not only wage goods but also health, education, etc, as well as the material cultures with which these are integral.

Interestingly, the more recent revitalized literature on social reproduction has examined how it is stratified by race and other forms of intersectional oppression as well as gender (Fine 2020a). From the SoP perspective, this suggests that the ways in which neoliberal forms of social reproduction are gendered or racialized are differentiated across sectoral (SoP-linked) practices that need to be unpicked in terms of the nature of sexism and racism themselves overall—as it were, such oppressions in housing and health differ from one another, even if mutually constitutive.[14]

The SoP approach, thus, has contributed to advances in understandings of consumption with particular regard to social policy (moving beyond its initial preoccupation with PSSoPs, especially around denationalizations). By doing so, the SoP approach has brought emphasis to the role played by the state in provisioning and reproduction as well as how it and other factors take on contemporary significance in the period of neoliberalism.

4.3 Operationalizing the SoP Approach: Social Reproduction in Practice

This section draws on a number of specific SoP case studies to give an illustration of how the approach has been applied in practice. The range of applications is extensive, from avocados (Serrano and Brooks 2019) to Zambian electricity (Bayliss and Pollen 2019) (and see Chap. 6 for more on the scope of SoP analyses). This section focuses on a handful of cases that have been selected to reflect some of the issues raised above regarding social policy and social reproduction. The sectors covered are housing, water, health services and fashion. Mostly these are from the UK but the chapter draws on case study material from South Africa in housing and water, and the fashion case study relates to global supply chains. These examples show that the materiality of what is provided, and the systems by which each of these reaches consumers or end users, are diverse in the extreme, across sectors and locations. Hence the drivers of consumption of housing or water compared with, say, fashion cannot be reduced to simple assumptions that are universally applicable. However, the studies cited below indicate that systemic structures run across individual sectors.

The cross cutting effects of global processes on the 'everyday' is a theme picked up in the rapidly expanding financialization literature, especially in light of (mortgage, credit card and other) indebtedness in accessing norms of consumption and managing daily life (Mader, Mertens & van der Zwan (Eds.) 2020). Accordingly, the presence of the everyday in consumption studies has been much more prominent with the transformation of consumption norms by financialization. It has become a major theme in current applications of the SoP approach—alongside globalization and neoliberalization, and their variegated forms.

The SoP approach is no exception in incorporating financialization as an element in understanding contemporary and everyday consumption, as illustrated by the discussion of housing below. But it also has a longer, pre-financialization record as it were, revealing other structures and social relations that underpin everyday consumption. And these we have to understand if we are to situate the impacts that financialization has wrought. With water in England and Wales, for example, turning on the tap is taken for granted. This is part and parcel of everyday life, unchanged over decades. But the mundane needs to be unpicked to find what lies beneath. Similarly SoP studies of other typical everyday activities, such as visiting the doctor or buying clothes, have highlighted major systemic shifts in provisioning with profound implications for cultures and understandings as well as for equity of outcomes. Such transitions in the underlying social relations are mostly out of view of the individual user or consumer, as discussed in more detail below.

4.3.1 Consumption Norms in Housing

As stated in Chaps. 1 and 3, the SoP approach has its origins in understanding consumption in terms of consumer norms, which are indicative of the propensity of certain groups differentially to own or to have access to certain (types) of goods or services.[15] The SoP approach is important for specifying norms of consumption not only in terms of who gets what, but how and why. Early SoP work was on consumption norms for consumer durables with many case studies (see Chap. 1). The notion of consumption norms is also well-illustrated by the housing system, where growing inequality of access and provision is heavily associated with the expansion of owner-occupation and mortgaging, and the decline of social housing (and even increasing role for private rented). A virtue of the SoP approach to housing has been for it to link changing forms of housing tenure to underlying forms of provision (the elements that make up housing supply) and how these are differentiated from one country to another. To a large extent, the housing literature has previously, and understandably, tended to emphasize the role of mortgaging and owner-occupation especially in light of sub-prime. But, increasingly, the housing literature has come to recognize the variegation in housing systems, a core insight from SoP analysis of housing. The increasing emphasis on the different incidence of mortgaging and of its impacts upon housing supply and access, confirms, if not necessarily explicitly deploying, one of the leading insights and operational aspects of the SoP approach, its emphasis on how housing SoPs (or others in their own sectors) are differentiated from one another even if sharing elements, such as land-dependence, construction and (financialized) mortagaging, in common as with housing, see Aalbers (2016, 2017) and Fernandez (2016).

Clearly, the way we consume (and produce) housing, and why, is different from the way we consume other things such as water, itself different from food and clothing etc. These differences are both obvious and subtle and may need spelling out as opposed to being left to the blunt explanatory instruments of supply and demand, price and income, tastes and preferences. For example, houses are (mostly) immobile. They have to be consumed where they are produced. While the share of income spent varies widely, in many cases housing takes up a substantial part of household expenditure. Housing standards are socially and historically determined, and they change over time and across locations. There are strong cultural associations with a home. It is not just the functionality of a house but its location and amenities that create its worth to its inhabitants. Houses are exchanged infrequently relative to other items of consumption, and they are long lasting with markets for new and old interacting with one another.

Some insights from analyses of SoP studies of housing follow.[16] Robertson's (2014, 2017) detailed study of the UK's housing SoP explores the dynamics that have given rise to the current dysfunctional structure, especially with increasing homelessness for some alongside immense housing wealth for others. The numbers of rough sleepers increased by 169% in England from 2010 and 2017.[17] Those in housing poverty in the UK are reported to have risen from 7.3% in 2012 to 12.5% in 2015, one in eight of the population, and even higher at one in seven in an urban environment. This is a dramatic indication of change in housing norms, who gets what and how, more on which below. How is this to be explained?

Robertson (2014) traces current housing patterns to a number of policies implemented by the Thatcher government of the 1980s which led to a shift in housing tenure norms. Financial deregulation, abundant liquidity and low interest rates led to the rapid expansion of mortgage lending from the 1980s in both the number of mortgage products and mortgages and the amounts being lent. This combined with the sale of social (local government rented) housing under the Right (of tenants) to Buy scheme in the 1980s and 1990s fuelled an increase in owner occupation.

Developments in housing consumption have, then, been underpinned by developments in the financial sector. Financial innovation, in the form of securitization of mortgage debt into bonds sold on international markets, meant that banks could lend without heeding capital requirements. Thus, there was greater penetration by finance into housing in the sense of an expansion in the number of people whose homes depended on mortgaging. Expansion of mortgage lending was associated with growing numbers of homeowners although the direction of causality is unclear (Robertson 2017).

The role of the state in housing in the UK was transformed by these reforms. Local councils were forced to sell social housing and their capacity to build social housing was withdrawn. The sale of social housing constituted a transformation of state welfare policy and was part of a bigger transformation discussed below. The state's retreat from direct production had an impact on the overall character of supply, much of which had been carried out previously under supervision of local authorities for provision of social housing. The neoliberal reforms led to an increase in the role of speculative house building firms in production, and the immediate consumers were no longer local authorities as potential landlords but end users or those purchasing new or existing stock to rent out and/or to accrue capital gains in a buoyant housing market.

The desirability of home ownership was, then, boosted by rapidly escalating house prices which resulted from a slow supply response to increasing demand. Housing supply consists of existing second-hand houses and newly built properties. The slow supply response is typically attributed to the combination of slow and opaque practices in government departments for land planning alongside housing developers retaining land for speculative purposes (known as 'land banking'). However, when viewed through the SoP lens, the supply response is a result of a combination of more and complex factors. While mortgage finance evolved rapidly in response to increasing demand for housing, and is rightly seen as a major factor in housing SoPs, less is known about the way that the financial sector operates in relation to housing production, especially relative to the study of mortgaging in the wake of the global financial crisis (GFC). There are different considerations around finance for large developers, who can access capital markets, than for smaller companies that are reliant on bank loans and whose financial sustainability depends particularly acutely upon appreciation in the value of the land upon which they are working. Shareholders in large companies also want them to maintain large land banks not just to profit from speculative gains but also as an indicator of long-term health and sustainability of earnings. 'The influx of mortgage finance had the effect of elevating the importance of house builders mercantile activities by increasing the premium attached to getting land trading right' (Robertson 2014, p. 46). This interacted with the planning system leading to poor supply responses overall. The relative importance of building firms land banking for speculative purposes, as opposed to the supposedly artificial unavailability of land for construction due to planning restrictions, remains a matter of dispute.

However, most of that dispute has taken place before the publication of Christopher's (2018) outstanding account of the privatization of ownership of land since Thatcher first came to power. He reports, despite claims of insufficient land for house building, that half of the UK's publicly owned land, 10% of the British land mass, has been sold to private interests! He observes that this land is worth 'somewhere in the region of £400bn in today's prices ... [it] dwarfs the value of all of Britain's other, better known, and often bitterly contested, privatizations'.[18]

The role of the state, at least in the UK case, in both sustaining land values and making it available for construction, points to tensions in how the state intervenes in housing systems, and how these may be resolved in differentiated ways, not least within the UK over time and the corresponding decline of council (and, as a potentially significant alternative) social housing.[19] Such tensions can even be played out between the different tiers of government. But what stands out, somewhat uniquely in the UK case, is the extent to which finance-fuelled house prices (primarily representing inflation of land values) have not been accompanied by a corresponding increase in housing supply. As a result, the housing bust associated with the global financial crisis has been less acute than elsewhere where collapses in house prices was accompanied by a collapse in supply. There has, though, been a common pattern across Europe of increase in housing benefit to support those who are unable to afford private rented housing (alongside lack of access to finance for owner-occupation and the decline in social housing as its funds have been shifted to housing benefit) (National Housing Federation 2017).

The result is basically state subsidy to private landlords, with private tenants acting as conduit between state and landlord. This has been most extreme in the UK. Indeed, in 2009, already around two-thirds (64%) of public spending on housing went into housing allowances (to subsidize renting) and 36% into building new homes. In 2015, the ratio had shifted even further towards housing allowances. In this year, 85% of all public spending on housing went into housing benefit and only 15% into building new homes.

In the UK, rising house prices, then, have had the effect of putting home ownership out of reach for many households and has led to the subsequent expansion of the private rented sector against previous trends. In 2015, 20% of households rented privately compared with 12% ten years previously. This too has been boosted by financial deregulation which led to expanded competition in financial services, allowing for the Buy to Let mortgage which, combined with growing value of housing equity, led more homeowners to buy houses for the purposes of letting these out and accruing capital gains on the purchase of their own homes as well as those of others to whom they rent privately. As a result, some homeowners have become more or less casual landlords, acting both as consumers and 'producers' of housing, although large-scale, global property companies have been major players in buying to let, attracted at least as much by rising values of their portfolios as the rents they might generate (especially in London, for example). The private rented sector had previously been characterized as a source of housing for younger, mobile members of the population but now provides more for low-income families with children and a growing number of elderly. It has the worst conditions and some of the most acute affordability pressures of all tenure types.[20]

Housing in the UK and elsewhere has understandably been dominated by a culture of the desirability of home ownership, as the costs and benefits attached to its provision (if accessible) far exceed those of (most) social or private renting (unless subject to negative equity, mortgage default and dispossession—rare in the UK context in absence of major bubble burst in housing markets). Owner occupation is depicted as a kind of innate instinct but Robertson (2017) demonstrates that this element of the material culture surrounding housing is relatively new and has been derived from material conditions surrounding access to housing and concerted campaigns by the state. As the housing market evolved and house prices appreciated, in light of the structural shifts outlined above, the individualized homeowner was encouraged to view a house less as a home and more as an asset. Neoliberalism has led to the creation of individual consumer entrepreneurs who accumulate assets to provide for their welfare. This was not a natural step but something that people had to be induced to do. Robertson (2017) considers why such narratives to normalize home ownership emanated from government. She finds that homeownership is seen as part of a broader neoliberal project of individualized asset-based accumulation to replace collectivized welfare provision, with a greater reliance on the market, encouraging individuals to bear more risk. Growth in owner occupation is crucial to this and relies on the intentional transformation of the culture of housing consumption and this was part of Thatcher's 'attempt to use housing to change the character of the individual and their role in the economy' (Robertson 2017, p. 403). The contemporary UK housing system has evolved in part in response to efforts of consecutive neoliberal governments to transform the welfare state from one centred around state provision to one where individuals bear more responsibility for their own welfare. Central to this is that the individual becomes an active consumer of financial assets. In a further twist, the resulting high proportion of personal wealth in housing now presents political pressures that stand in the way of redistributive policies.

Such cultures are, however, contradictory. The house as a home is associated with stability, constancy and home making whereas the house as financial asset is linked to speculation and climbing up the property ladder. Robertson highlights the chaos resulting from such housing cultures as extremes emerge such as empty properties in London that have been bought solely for the purpose of financial speculation. As house prices have escalated, there is a collective experience of exclusion from owner occupation under the rubric 'generation rent' (Robertson 2017, p. 405). But the bedrock desire for home ownership persists. It gives rise to contestation around housing cultures with a new discourse of housing as a socioeconomic right. The system of housing provision in which owner occupation is dominant is increasingly dysfunctional inciting resistance and putting pressure on, and struggles around, current housing norms (Robertson 2017).

Isaacs' (2015) study focuses on a very different housing SoP—for low cost housing, and associated policies, in South Africa. In principle, if not in practice, these are designed to address the housing needs of poorer households in the wake of the inequalities of access and provision inherited from the apartheid era. As in the UK, home ownership was depicted as a source of both shelter and a financial asset. The increase of both housing consumption and production took a clear neoliberal form. The approach to consumption has sought to turn houses into financial and economic assets in addition to providing shelter, with a focus on individual ownership of freestanding houses. Housing has been portrayed as an asset and household mobility to be achieved through the 'property ladder'.

Housing construction in South Africa was outsourced to large private sector construction firms funded via capital subsidies. The land that was developed for housing was on the urban periphery far from economic infrastructure and employment, reproducing apartheid's spatial features. The rhetoric was of state support for private sector delivery, with household access reliant on borrowing from private finance. However, the policy has been constrained by the banks being unwilling to lend to low cost housing projects and low-income households. Households have sought various means to access credit including use of high interest informal sector loans. Finance and indebtedness have become more entrenched in daily lives as those on low incomes, heavy social security dependence and high unemployment seek access to housing and subsistence more generally.

Isaacs shows how housing policy failed to adequately address the intersecting horizontal factors shaping deprivation including poverty, unemployment, the impact of the HIV/AIDS pandemic, poor education and a lack of understanding of housing issues amongst borrowers and many others. These have contributed to a downward spiral where houses are not adequately maintained, rates and charges are unpaid to local authorities which are then unable to maintain public facilities. As a result, property values fall and institutional and individual investors leave the neighbourhood. Owners are then left with property worth much less than they paid for it, and the nirvana of a property-owning democracy remains elusive.

Isaacs documents the tensions between the different levels of government and the private sector in ways which ultimately lead to what he describes as 'business as usual: the marginalization of communities, poor delivery and large capital involving itself only in the most lucrative aspects of the SoP while shunning the rest' (Isaacs 2015, p. 133). While the notion of housing as an asset has entered policy discourses, the approach is inoperable within the putative low-cost housing market.

These examples highlight the insights that can be derived from the application of the SoP approach to the drivers of consumption norms in housing. They emphasize the importance of context, demonstrating the contrasting outcomes from the promotion of systems of housing access mediated by private credit finance. Consumption outcomes result from the intersection of numerous elements of the SoP. Consumer choices themselves do not emerge spontaneously out of some kind of innate set of preferences but are malleable and shaped by those that have an interest in certain outcomes.

Significantly, the case study of housing for South Africa marks a major shift in neoliberal thinking and practice with, in the context of development, World Bank scholarship, ideology and policy in practice to the fore, in promoting 'market' delivery, that is, state policies in support of private provision and financing.[21] For preparations for a post-apartheid society in the early 1990s, it was anticipated that prospective policies would redress some of the inherited inequalities through a major programme of state-led social provisioning. Indeed, with one of us serving as part of a team formulating policy for the African National Congress (ANC) Government-in-waiting in South Africa, the SoP approach was in mind, if not explicitly mentioned, in housing (and other social) policies being formulated for the ANC just before the end of the apartheid era. However, once the ANC came to power, these policies, of large-scale provision of social housing for rent, alongside other socially progressive measures such as for more equitable education and health, and electricity access, were discarded as neoliberal stances came to the fore in government.[22]

4.3.2 *Water*

Like housing, water has material features that distinguish its production and consumption from those of other commodities. A key element of water is its leakiness. It seeps out of pipes and taps and it evaporates. It flows downhill and can be difficult to contain, and leaks are often difficult to detect. Water sources sometimes have to be shared across regional and international boundaries. It is heavy to transport relative to value and so is typically used close to source. Delivery systems tend to be fixed and capital-intensive, relying on networks of pipes and pumps that are not easily moveable, so investments are long-term. There are considerable scale economies, and delivery is usually monopolistic. The availability of water is unpredictable due to variability in rainfall and is also affected by pollution and climate change.

In England and Wales (EW), most people access water via household taps and have little regard to the origins of their water. Their main interaction with the production process is just through the payment of water bills. As far as the consumer is concerned, little appears to have changed since the sector was privatized in the 1980s. However, this superficial continuity conceals substantial changes in the underlying social and economic relations in water provisioning. Bayliss (2014, 2017) found in her analysis of the SoP for water in EW that, while for consumers, water flows through the tap as always, the nature of producers, and their methods of maximizing shareholder returns, had changed dramatically.

The water sector in EW was privatized in 1989 when the regional public water companies were listed on the London Stock Exchange. Water is now provided by ten integrated, privately owned, regional water companies (and some smaller water only companies), each responsible for the whole chain of activities from abstraction to water treatment and distribution, delivery to end users, billing and waste-water management. Bayliss (2014) showed that the ways that these companies operated varied, based on the nature of their ownerships. She found that the four companies that were owned by consortia of financial investors (mostly based offshore in Jersey) had distinctive corporate structures that were different from those of other ownership types such as those that remained listed on the stock exchange.

Bayliss found that the reason for this was that these finance-owned companies had been able to use complex securitization structures via Cayman Islands subsidiaries to increase their debt levels substantially. With a complex vertical chain of intermediate holding companies, interest and dividend payments were channelled to other group companies in a series of opaque transactions. A portion of the debt raised was used to pay special dividends to shareholders and to pay for the acquisition debt (the cost to the investors of acquiring the utility), which was added to the debts of the water company. As a result, some water companies became heavily indebted. Investors also make profits in other ways that are unrelated to production processes. Methods include loans from shareholders to the water utility at higher than market rates of interest, the sale of company land and reservoirs and in the selling of segments of company ownership, for prices that are undisclosed. Financing costs are paid from consumer bills. Thus, the SoP analysis found that while consumption practices had changed little, there had been a major change in the substance of what was being financed.

Water has essential health and social properties and provision is monopolistic so interactions between consumers and producers are mediated in EW by the state regulator, Ofwat. Regulation largely takes place via a system of price controls. Prices are fixed in advance every five years based on a negotiated assessment of investment needs, estimated costs and performance against targets in the previous five years. The system is intended to mimic competition with firms supposedly operating as price takers as they would if under conditions of perfect competition. Ofwat is also required to increase competition wherever this is possible

While depicted as an independent technocratic exercise, the rules of regulation are complex and fluid. The price setting process is heavily contested. A number of price reviews were considered to be generous to investors, mainly in the way that the actual cost of debt was consistently below what was assumed in price setting (PAC 2015). The regulator makes choices about where and how to intervene. For example, the regulator does not intervene in company debt levels and dividend payments as these are deemed to be financial 'market outcomes', subject either to its own competition or regulation (and so somebody else's business). However, fluctuations in demand (also arguably a market outcome) are an area for intervention. Companies are compensated for falling demand by an increase in the price they can charge in the subsequent price review process. Consumers are encouraged to save money by lowering consumption, but they are not told that it will lead to a price rise in the future. With the sector conceived as if governed by a market, access is the responsibility of individuals. Greater metering of water consumption puts the onus on households to ensure affordability by reducing consumption creating new 'financial subjectivities within the home' (Loftus, March & Nash 2016, p. 329).

Companies are required to ensure that they retain a credit rating that is 'investment grade'. This acts as a brake on their borrowings. But it creates complexities for regulation as, in practice, the relations between regulator and regulated are complicated by the fact that the regulatory framework itself has a major impact on the perceived creditworthiness of the companies. The regulatory rationale of protecting the interests of consumers is inherently compromised by the need for the sector to sustain investor-friendliness. Moreover, the ambition of transparency in regulatory processes is lost in the density of activity that has become involved. Regulation is incredibly complex. The documentation for the last price review, PR19 led to hundreds of reports and some of these ran to hundreds of pages. The shift to a supposed market is highly interventionist, with the state intervening selectively. As water flows one way with a turn of the tap, finance flows in the opposite direction, from households to (offshore) global financial capital.

The lack of popular/consumer dissent around water privatization and company profiteering is striking. Until nationalization of water reached the manifesto of the (since defeated) opposition Labour Party in 2017, there was barely a murmur of protest while other parts of the world have rioted against far milder forms of privatization. Bayliss (2017) considers how cultures have been constructed and construed around water provision and consumption so that such profiteering has become not just tolerated but normalized through the narratives of efficiency, investment and market outcomes. The state plays a key role in the neutering of contestation in the water sector, notably in conflating the interests of private shareholders with those of water consumers. What is best for companies is also considered to be best for customers. A discourse of harmony and shared interest conforming with commodity provision, is projected while potential conflicts between agents are overlooked.

Regulation is depicted as a technical exercise so that the political implications of the distributional outcomes from regulatory processes are ignored. An important narrative that supports the financialized cultures of water privatization stems from the understanding of the sector as a (imperfect) market and this is embedded in regulatory practice. Consumers are invited via different regulatory procedures to express their views on sector performance as if they were able to exercise choice, and customer satisfaction is a metric in assessing water company performance. However Bayliss (2017) cites evidence to indicate that such views are malleable, affected by advertising, and the relative levels of water bills compared with other costs. Thus consumer views are moulded to meet the performance metrics. The process of opinion gathering is selective so that consumers may be asked if they would recommend their water company to friends but they are not asked their opinions on bloated dividends or directors' pay. Contestation has been channelled along the lines of narrow superficial consumer consultations, and regulation is skewed to address specific issues such as capital costs, leakage and customer service. Meanwhile the securitization of water bills and the extraction of revenue by private equity investors is unquestioned.

This SoP investigation is revealing of how global finance drills down to everyday activities, in ways of which consumers are mostly unaware. This is not, however, a generic feature of water systems. It is reflective of the specific combination of water in the context of EW. No other country has implemented such profound privatization. The significance of context is highlighted in a contrasting study of the SoP for water in South Africa. Here, as with EW, Bayliss (2016a) maps the agents involved in the production and consumption of water as well as the different state agents involved.

There are fundamental differences in the context of the SoP for water in South Africa, compared with EW. For example, unlike EW which, at the time of writing in February 2020, was facing unprecedented flooding, South Africa is highly water stressed; the economy is dominated by the water-intensive mining industry which is a major contributor to water quality degradation, and the country has a legacy of high and racialized inequality in access and provision following the apartheid era. These and other contextual differences mean that the water SoP in South Africa takes a different form, and raises different concerns from that in EW.

In contrast to EW, there is very little private sector involvement, and the supply of water in South Africa is stratified according to function. Raw water is abstracted from surface or ground water which may go directly to the consumer if they can use untreated water (such as irrigated agriculture or the mining sector), or it goes to regional Water Boards. These treat and transport water to some (large-scale) end users or to Water Service Authorities, many of which are municipalities that provide the water to end users. This stratified approach to the water SoP in part, then, means that consumption is organized according to user type rather than according to location. The result has been the construction of infrastructure networks that bypass communities and rural households so that wealthy mineral production sits next to shack housing where residents lack basic services.

For households, a major post-apartheid feature has been the extension of access to low-income consumers in informal settlements. As in EW, the state is responsible for protecting the interests of consumers. And, as with housing, the provision of water in South Africa highlights the contradictions of progressive policy rhetoric that intersects with neoliberal practices. The country was one of the first to guarantee a basic amount of water to its citizens with the Right to Water enshrined in the country's 1996 Constitution. The roll out of infrastructure provided the potential for a dramatic increase in water access. But the widespread use of prepaid meters and a tariff structure which escalated rapidly for consumption above the basic amount, penalized households with large numbers or those with shared connections, each of which tends to be the poorest. Many have had their supplies disconnected and struggles around water have been widespread (Bond 2014; Angel and Loftus 2019). Rather than promoting equitable access, for some the Constitution and the basic lifeline helped to deflect criticism and provided something of a 'moral justification for disconnections' (Loftus 2004, p. 195). For Bond (2014, p. 462), rather than protecting the marginalized, the Constitution facilitates inequality as it 'serves as a mythmaking, deradicalizing meme'.

The comparison of the SoPs for water between South Africa and EW highlights the way that the act of consuming water, an essential commodity which is largely homogenous, subject to variations in quality and access, is to engage with a complex set of social relations that are highly variegated. In both cases, a superficial technical neutrality conceals a bias in favour of the economically and politically powerful—the financial sector in EW and the mining sector in South Africa. The state in both cases plays complex and contradictory roles with lip service to social policy while at the same time supportive of the extractive practices of the neoliberal machinery (Bayliss 2016b).

4.3.3 *Health Services*

Unlike water or housing, health services are highly labour-intensive. In England, as with water, health services have changed little from the perspective of consumers. Access remains free at the point of delivery but provisioning has been transformed in all aspects of the SoP, including the agents, structures, processes and cultures. Bayliss (2016c) examines the SoP for health services in England, although it is acknowledged that the factors that combine to create healthy people (or otherwise) are extensive and go far beyond the provision and consumption of services per se. Bayliss shows that, as with housing and water, the National Health Service (NHS) has undergone numerous neoliberal restructurings since the 1980s. Successive reforms have been introduced to make the different public sector elements of the health system behave more as if they were private sector entities. This began in the early 1990s with the introduction of 'internal markets' which made a distinction between healthcare 'providers' and 'purchasers', and numerous subsequent interventions have sought to tweak this model. Since 2004, health providers have been remunerated according to a complex 'pricing' system known as 'payment by results'. Further market-like interventions include the introduction of 'patient choice' in the 2000s, which allowed for private providers funded by the NHS. In addition, private 'Independent Sector Treatment Centres' were introduced in private hospitals to provide routine treatments to increase choice and to clear NHS waiting lists. The role of the private sector expanded further following the 2012 Health and Social Care Act with two key interventions. First, state health providers can earn up to half of their income from private patients. Previously the limit was set at 2%. And, second, contracts had to be put out to tender, to Any Qualified Provider, paving the way for the private sector to provide state-funded health services at a much greater scale than previously. The role of non-NHS providers in NHS-funded services has increased markedly.

There is little evidence of greater efficiency as a result of these market-oriented interventions.[23] While progress may have been made with some procedures, the opposite can hold for health and care services as a whole with the passing on of burdens. Following the introduction of payment by results, there is evidence of a reduction in length of hospital stays, and an increase in the proportion of day cases but no evidence of change in care quality. However, providers have limited incentives to integrate treatments, innovate in delivering care to people with long-term health conditions, and shift patients from hospitals to community settings. Providers may also have incentives to over-treat or categorize patients as having more severe conditions than they actually have ('upcoding'), discharge patients too early, and focus on lower-cost patients.

As with other SoP studies, the context is crucial for understanding the outcomes from the system for health services. Reforms in England are taking place against a backdrop of austerity. This has not just affected health spending itself but health services are also suffering due to the knock on effects of cuts to social care. The sector has been crushed by an enormous deficit. Health providers are striving to cut costs with unrealistic targets set in the pressure for 'efficiency savings'. A 1% cap (since lifted) on pay increases for health sector workers was imposed in 2010, leading to a reported fall in real wages by 15% by 2017. Health services are largely labour-intensive. Labour costs account for 63% of total expenditure (Bayliss, Fine, Robertson & Saad Filho 2020). The pay freeze has been counterproductive leading to staff shortages which are compensated for by using much costlier temporary agency staff. Many hospital trusts are running at capacities that far exceed recommended levels.

There has, though, for the past decade been a narrative that presents the NHS as unaffordable and inefficient, despite evidence that puts health costs and performance at about average for OECD countries. The backdrop of austerity gives additional weight to these narratives of inefficiency, and the depiction of the public health service as being in crisis increases the attractiveness of private healthcare providers. Segments of health provision now feature in the investment portfolios of transnational conglomerates not normally associated with the sector. For example, Virgin Care (part of Virgin Group Holdings Ltd, incorporated in the British Virgin Islands) has contracts for the provision of Child and Adolescent Mental Health and for and social care services (Bayliss 2016c).

The SoP approach in this case highlighted the transformation of health services over the past three decades in ways that are not apparent to the patient. Finance has taken greater prominence, with private finance and provision playing a greater role and public sector entities facing financial penalties for failing to meet targets which, in many respects, are outside of their control. This section gives a brief overview of some of the complexities and contestations among the agents involved and the way that these shift with commodification. There are numerous other ways in which the commodification of health services has shifted the social relations of provisioning.[24]

4.3.4 Clothing and Fashion

Clothing is also an item used (if not exactly consumed) by all of us every day. The production of clothing encompasses an extensive SoP with agents typically reaching across continents. The chain of provisioning incorporates agricultural workers and machinists usually from the global South with retailers, designers, models and advertisers often in the North. Brooks (2015) draws on the SoP approach to show how clothes are constructed through a web of relationships which draw upon and reproduce positions of power and dependency emerging from decades, even centuries, of uneven development (p. 35).

'Fast' fashion, defined as the way that trends move rapidly from the catwalk to the high street, results from extensive and complex networks of international trade. In contrast to the SoPs discussed above, the production of clothing is far more flexible in how it is internationally organized. Cotton is produced in one country, cut and sewn in another, and sold in another. The quest for cheap but quality clothing has led to the migration of industries to the less expensive South and the depression of wage rates for many workers. African agricultural labour receives only a tiny proportion of the final sale price. The distribution of profits is vastly unequal with the owners of global fashion brands amassing billions of dollars in fortunes while countless garment workers earn a few dollars a day or less. Labour standards are often low for agricultural workers and cotton ginneries. Workers have few rights and fast fashion factories can easily close and reopen elsewhere in the event of labour protests. Conditions in the South are exacerbated by environmental impacts. Cotton production is water-intensive and often grown in water-scarce contexts.

Brooks (2015) carefully traces the historical evolution of both global clothing production systems and fashion systems from ancient civilizations through the industrial revolution to the factory-based SoPs of today, each system facilitating exploitation of workers at one end of the chain so consumers can be fashionable at the other. He shows that the global fast fashion industry, while spanning countries, derives from the workings of this historical trajectory. It has created today's specific patterns of consumption and production.

Consumer preferences are directly and indirectly shaped by producers. Material cultures around fast fashion are strongly geared towards ever-expanding consumption. Fashion firms have created opportunities for sales growth by increasing the speed at which tastes and fashion change. When it comes to clothing, for the most part, 'value' is socially determined and not related to usefulness as such. Creating desirability and cultural attachment to items of clothing via advertising and other means are central to value creation. Brooks charts the cultures associated with denim jeans by way of example, from hardwearing work wear, especially of the iconic cowboy, through youth culture to default wardrobe item. Jeans have become popular because of their ease of manufacturing and in meeting practical and cultural needs. But there is considerable differentiation across styles. More generally, fashion cultures are shaped by well-paid designers, advertisers and fashionistas. Aspirational images compel customers to buy more clothes, far in excess of need. The fashion industry creates financial and psychological pressures simultaneously (and contradictorily) to create an individual identity as well as to subscribe to current norms, giving rise to what Brooks terms a 'consumption arms race' (p. 33). Meanwhile, the relations between consumers and the producers are obscured. Brooks' study of the clothing industry links high street shoppers in the North (and the wealthy in developing regions) with (often impoverished) agricultural workers and even child labour in cotton production in the global South.

The unequal relations that are embedded in the production of cotton, which is an input into the global clothing industry, dates back to the colonial era. More recently structural adjustment policies in the 1980s that put a 'naive faith in the magic of the market' (Brooks 2015, p. 112) led to the removal of state marketing boards and exposed farmers to volatilities in global agricultural markets. These policies were advocated by policy advisers in the global North, such as the World Bank who viewed price controls as a form of 'market distortion'. Yet the subsidies paid to cotton farmers in the USA continued. Brooks traces this unequal engagement in international trade to colonial exploitation which set the stage for the development of modern capitalism. Domestic clothing and textile industries in African nations have struggled to compete with production elsewhere, particularly as China has become the dominant player in clothing production.

Brooks notes a further dimension to the SoP for fast fashion has been the 'overconsumption' of clothing as reflected in the vast surplus that is donated to charity and which makes its way via an opaque 'shadow trade' to developing countries where it is sold or dumped. The agents involved in such trading are difficult to trace but low-income countries are consigned to the cast offs of the wealthy and are subtly connected to the vagaries of the international clothing industry. This second-hand trading has created another layer of relations and cultures which, as with SoPs more generally, are mostly hidden from view. The social relations behind the donated clothes are not displayed. This importing of used clothing has contributed to the demise of domestic garment industries.[25] However, cultural associations are complex and diverse. Hansen (1999, 2000) draws on the SoP approach to explore the second-hand clothing trade in Zambia. She links the export trade with the local consumption practices that have arisen around it, describing the SoP approach as a 'rich organizational frame' for her research (p. 17). She shows how imported clothing from the west has acquired cultural meanings and, in some cases, is highly prized and deemed to be 'among the prerequisites of modern society' (Hansen 1999, p. 358). She sees practices around the consumption of such clothing as contributing to a broader sense of post-colonial society and the location of Zambians within this.

4.4 Conclusion

The above review of case studies shows how the SoP approach provides a systematic framework to manage the extensive and multifarious factors that shape provisioning outcomes. The studies show that individual preferences play only a small part in determining who gets what. Production and consumption do not take the form of a market tending towards some equilibrium. Rather, outcomes emerge from contested social relations embedded in longstanding structures and processes. The approach goes beyond simple tracking of value chains and agents, to consider how these interact and how the state is involved in promoting specific outcomes.

In each of the studies the consumer is distanced from the social relations that underpin provisioning in policy and mainstream narratives. In housing, water, health services, fashion and food (Chap. 5), the systemic functioning of these systems are obscured from the consumer. Financialization in housing and water is located at considerable distance from those taking out mortgages let alone using their taps. Similarly, the attention of the consumer is to the fashion item as such rather than to the social relations by which their provisioning is sustained or to the side industry of the on-sale of second-hand clothes to the global South. This is hidden from those that donate used clothing.

Such distancings or, more exactly, concealments serve to focus attention on the commodity and on the capacities of the consuming individual. This ties in with a culture of self-responsibility, whether supported, subsidized or otherwise. The narratives in both housing and water in the UK are such that the individual is responsible for ensuring access and affordability, ostensibly in isolation from the underlying structures (and revenue extractions therein) of production. This is part of broader narratives and neoliberal cultures which promote individual and individualized responsibility which are typically at the expense of the collective, across ever expanding terrains of economic and social life.

When the SoP approach lifts the lid on the production system, and its connections to consumption, the ways that agents are competing to capture value in the system is revealed. The SoP approach draws attention to the conflicting priorities of agents in provisioning. These often take place out of sight of the consumer embedded in corresponding cultures that fail to question the extraction of revenues and the consequences of the dependence of provisioning on doing so. Putting the slogan 'People before

Profits' potentially runs extremely wide and deep in both principle and practice!

Notes

1. See Bayliss and Fine (Eds.) (2008) for the putative rethink on privatization by the World Bank.
2. For economics imperialism, see Fine and Milonakis (2009) and, most recently in context of interdisciplinarity, Fine (2019).
3. On Marcuse as the theorist par excellence of one-dimensioning (not one-dimensional) man, see Fine (2017a).
4. For the distinction between horizontal and vertical (in the context of consumption), see earlier chapters. In brief, the vertical follows provisioning from production through to consumption for specific goods, whereas the horizontal emphasizes cross-society factors such as utility, emulation, distinction, psychological factors, gender and so on.
5. For critique of mainstream economics on consumption, see Fine (1996a, 1998). Further, sociological concepts such as conspicuous consumption, emulation and distinction tend to be overgeneralized and presence needs to both commodity- and context-specific and rooted in the conditions of provisioning and corresponding cultures. See Fine (1994) on Veblen for example. Fine and Leopold (1993) and Fine (2002) critically synthesize from across the social sciences.
6. Trickle-down is another concept that has been criticised from the SoP perspective, and even contrasted with trickle-up as with, for example, denim jeans—from workwear to fashion (see below), with elites not necessarily leading fashion nor sources of demand.
7. In the USA, there has been an interesting inversion of the meaning of consumerism, inspired by Ralph Nader—from hedonism to activism.
8. For a child in apartheid South Africa, it was famously 'two dogs and freedom'.
9. To some degree, the preoccupation with false/true needs and the trope of achieving satisfaction or more in alienated contemporary capitalism has given way to 'happiness' research. This has been broached through the SoP approach, especially in addressing critically the finding that reported happiness did not tend to decline over the course of the global financial crisis. This is seen to be a consequence of heavily reduced aspirations as the target for achieving happiness, the analytical remedy to which is to examine how material culture of provisioning governs what is provided and how it is perceived through the SoP prism. See Boffo, Brown & Spencer (2013, 2017) and also Bayliss, Churchill, Fine & Robertson (2016).

10. For an account, drawing upon the SoP approach, and its 10Cs, of how ethics might be constructed in critical response to mainstream economics, see Fine (2013).
11. See earlier discussion of housing where what can be meant by entitlement/capability is irreducibly attached to forms of tenure and how they are provided. For corresponding critique of the entitlement approach to famines in part from a SoP perspective, see Fine (1997). And see Gough (2019) in bringing together the role that might be played by the SoP approach, together with ethics, sustainability and capabilities, in providing for universal basic services.
12. See Fine (2012, 2014, 2016, 2017b). Note that the SoP approach has been influential in research programmes on social policy for United Nations Research Institute for Social Development (UNRISD)—synthesizing the developmental welfare state with the SoP approach, Fine (2016)—as well as elsewhere, in education for example and in understanding the incidence and impact of PPPs, Fine (2020b), and for economic infrastructure in the iBuild project, http://sure-infrastructure.leeds.ac.uk/ibuild/ The SoP approach even found its way into the UK Treasury's way of valuing infrastructure (to support value added other than through the market), in reference to systems of infrastructure provision, see HM Treasury (2015, 11).
13. For an account drawing upon the SoP approach, with some emphasis upon how to understand the 'moral historical' determination of the value of labour power, see Fine (2020a).
14. And the same applies to provisioning by age and disability (Fine 2018), with disability in particular at the forefront of austerity measures. See also Ryan (2019).
15. Empirical investigations of such shifting (gendered) norms began in the SoP approach with consumer durables in light of their putative impact on (female) labour market participation covering, for example, cars, washing machines, central heating, dishwashers, videos, microwaves, telephones, and so on, see Chap. 1.
16. These were part of a broader comparative FESSUD study of housing SoPs also including Portugal, Poland and Turkey, see Robertson (2016).
17. See Wilson and Barton (2020).
18. See https://www.theguardian.com/commentisfree/2018/feb/08/biggest-privatisation-land-margaret-thatcher-britain-housing-crisis and also Christophers (2018).
19. A SoP-inspired study of Turkish housing, for example, reveals the facilitating role played by the state in stimulating both public and private housing provision, see Unsal (2017).
20. See Wilson and Barton (2020).

21. For the evolving relationship across World Bank scholarship, ideology and policy—through Washington and post-Washington Consensus and beyond—see Bayliss, Fine & Van Waeyenberge (Eds.) (2011) and Fine, Johnson, Santos & Van Waeyenberge (2016). On housing in particular, see Van Waeyenberge (2018).
22. See MERG (1993), which was commissioned in 1992 by Nelson Mandela but ruthlessly abandoned by the end of 1993, and Padayachee and van Niekerk (2019) for a full account of the abrupt discarding of the policies formulated in the ANC's Freedom Charter and beyond. The framing of social and economic infrastructure provision in the MERG Report through the SoP approach was made explicit in a draft prepared for a policymaking workshop to be held in June, 1996, but it was not delivered. See Fine (1996b) for text of the presentation.
23. For the absence of any evidence that the new public sector management has delivered any greater efficiency, see Hood and Dixon (2015).
24. This section gives a brief overview of developments in the SoP for health services in England. Privatization and financialization are evident in numerous aspects of health services, including in hospital construction and maintenance via the private finance initiative, in GP and dental practices and in care homes. See Bayliss, Fine, Robertson & Saad Filho (2020) for more details.
25. Some African countries attempted to ban the import of second-hand clothes in an effort to strengthen their domestic industries. In the case of Rwanda, this has escalated into a trade battle with the USA (John 2018).

References

Aalbers, M. B. (2016). *The financialization of housing: A political economy approach*. London: Routledge.

Aalbers, M. B. (2017). The variegated financialization of housing. *International Journal of Urban and Regional Research, 41*(4), 542–554.

Angel, J., & Loftus, A. (2019). With-against-and-beyond the human right to water. *Geoforum, 98*, 206–213.

Bayliss, K. (2014). *The financialisation of water in England and Wales*. FESSUD Working Paper Series, No. 52. Retrieved from http://fessud.eu/wp-content/uploads/2015/03/Case-study-the-financialisation-of-Water-in-England-and-Wales-.

Bayliss, K. (2016a). *Neoliberalised water in South Africa*. FESSUD Working Paper Series, No. 204. Retrieved from http://fessud.eu/wp-content/uploads/2015/03/FESSUD_WP204_Neoliberalised-Water-in-South-Africa.pdf.

Bayliss, K. (2016b). *The system of provision for water in selected case study countries.* FESSUD Working Paper Series, No. 194. Retrieved from http://fessud.eu/wp-content/uploads/2015/03/FESSUD.

Bayliss, K. (2016c). *The financialisation of health in England: Lessons from the water sector.* FESSUD Working Paper Series, No. 13. Retrieved from http://fessud.eu/wp-content/uploads/2015/03/Financilisation_Health_England_WorkingPaper131.pdf.

Bayliss, K. (2017). Material cultures of water financialisation in England and Wales. *New Political Economy, 22*(4), 383–397.

Bayliss, K., Churchill, J., Fine, B., & Robertson, M. (2016). *Summary report on the impacts of financialisation and of the financial crisis on household well-being.* FESSUD Working Paper Series, No. 199. Retrieved from http://fessud.eu/wp-content/uploads/2015/03/FESSUD_WP199_Summary-report-on-the-impacts-of-financialisationfinancial-crisis-on-household-well-being.pdf

Bayliss, K., & Fine, B. (Eds.). (2008). *Privatization and alternative public sector reform in sub-Saharan Africa: Delivering on electricity and water.* Basingstoke: Palgrave Macmillan.

Bayliss, K., Fine, B., Robertson, M., & Saad Filho, A. (2020). *Neoliberalism, financialisation and welfare: The political economy of social provision in the UK.* Cheltenham: Edward Elgar, in preparation.

Bayliss, K., Fine, B., & Van Waeyenberge, E. (Eds.). (2011). *The political economy of development: The World Bank, neoliberalism and development research.* London: Pluto.

Bayliss, K. & Pollen, G. (2019). *From global to local: Exploring the effects of the global energy policy paradigm in Zambia.* Sustainability Research Institute Working Paper Series, No. 119, University of Leeds. Retrieved from https://sri-working-papers.leeds.ac.uk/wp-content/uploads/sites/67/2019/10/SRIPs-119.pdf.

Boffo, M., Brown A., & Spencer D. (2013). Has the financial crisis really improved well-being? A critique of "Happiness Economics" and reassessment of the 2013 World Happiness Report. In A. Brown, D. Spencer, B. Fine, & A. Santos (Eds.), *Report on financialisation and well-being.* Leeds, UK: FESSUD Project, (Deliverable D5.01).

Boffo, M., Brown, A., & Spencer, D. (2017). From happiness to social provisioning: Addressing well-being in times of crisis. *New Political Economy, 22*(4), 450–462.

Bond, P. (2014). Constitutionalism as a barrier to the resolution of widespread community rebellions in South Africa. *Politikon, 41*(3), 461–482.

Brooks, A. (2015). *Clothing poverty: The hidden world of fast fashion and second-hand clothes.* London: Zed Books.

Christophers, B. (2018). *The new enclosure: The appropriation of public land in neoliberal Britain*. London: Verso.

Fernandez, R. (2016). *Stylized facts from housing and finance: How do they relate across space and time?* REFCOM Occasional Working Paper. Leuven: KU Leuven. Retrieved from http://www.fingeo.net/wordpress/wp-content/uploads/2017/07/WP1_Stylized-facts-from-housing-and-finance-1.pdf.

Fine, B. (1994). Consumption in contemporary capitalism: Beyond Marx and Veblen—A comment. *Review of Social Economy, LII*(3), 391–396.

Fine, B. (1996a). From political economy to consumption. In D. Miller (Ed.), *Acknowledging consumption* (pp. 127–163). London: Routledge.

Fine, B. (1996b). Some perspectives on the provision of social and economic infrastructure. Draft for proposed workshop for South African Policy Makers, June. Retrieved from https://eprints.soas.ac.uk/32961/.

Fine, B. (1997). Entitlement failure? *Development and Change, 28*(4), 617–647.

Fine, B. (1998). The triumph of economics: Or 'rationality' can be dangerous to your reasoning. In J. Carrier & D. Miller (Eds.), *Virtualism: The new political economy* (pp. 49–73). Oxford: Berg.

Fine, B. (2002). *The world of consumption*. London: Routledge.

Fine, B. (2012). Financialisation and social policy. In P. Utting, S. Razavi, & R. Buchholz (Eds.), *Global crisis and transformative social change* (pp. 103–122). London: Palgrave Macmillan.

Fine, B. (2013). *Economics—Unfit for purpose: The director's cut*. SOAS Department of Economics Working Paper Series, No. 176. Retrieved from http://www.soas.ac.uk/economics/research/workingpapers/file81476.pdf. Revised and shortened to appear as, Economics: Unfit for purpose. *Review of Social Economy, LXXI*(3), 373–389.

Fine, B. (2014). *The continuing enigmas of social policy*. Paper prepared for the UNRISD project: Towards universal social security in emerging economies, UNRISD Working Paper 2014-10. Retrieved from http://www.unrisd.org/Fine.

Fine, B. (2016). The systemic failings in framing neo-liberal social policy. In T. Subaset (Ed.), *The great financial meltdown: Systemic, conjunctural or policy created?* (pp. 159–177). Cheltenham: Edward Elgar.

Fine, B. (2017a). From one-dimensional man to one-dimensions economy and economics. *Radical Philosophy Review, 20*(1), 49–74.

Fine, B. (2017b). The continuing enigmas of social policy. In I. Ye (Ed.), *Towards universal health care in emerging economies: Opportunities and challenges* (pp. 29–60). London: Palgrave Macmillan.

Fine, B. (2018). Collective choice and social welfare: Economics imperialism in action and inaction. *Ethics and Social Welfare, 12*(4), 393–399.

Fine, B. (2019). Economics and interdisciplinarity: One step forward, N steps back? *Revista Crítica de Ciências Sociais, 119*, 131–148.

Fine, B. (2020a). Framing social reproduction in the age of financialisation. In A. Santos & N. Teles (Eds.), *Financialisation in the European periphery: Work and social reproduction in Portugal*. London: Routledge.

Fine, B. (2020b). Situating PPPs. In J. Gideon & E. Unterhalter (Eds.), *Critical reflections on public private partnerships*. London: Routledge, forthcoming.

Fine, B., Johnson, D., Santos, A., & Van Waeyenberge, E. (2016). Nudging or fudging: The World Development Report 2015. *Development and Change, 47*(4), 640–663.

Fine, B., & Leopold, E. (1990). Consumerism and the industrial revolution. *Social History, 15*(2), 151–179.

Fine, B., & Leopold, E. (1993). *The world of consumption*. London: Routledge.

Fine, B., & Milonakis, D. (2009). *From political economy to freakonomics: Method, the social and the historical in the evolution of economic theory*. London: Routledge.

Gough, I. (2019). Universal basic services: A theoretical and moral framework. *The Political Quarterly, 90*(3), 534–542.

Hansen, K. (1999). Second-hand clothing encounters in Zambia: Global discourses, Western commodities, and local histories. *Africa: Journal of the International African Institute, 69*(3), 343–365.

Hansen, K. (2000). *Saluala: The world of secondhand clothing and Zambia*. Chicago: Chicago University Press.

HM Treasury. (2015). Valuing infrastructure spend: Supplementary guidance to the Green Book. Crown copyright, Retrieved from PU1798_Valuing_Infrastructure_Spend_-_lastest_draft.pdf.

Hood, C., & Dixon, R. (2015). *A government that worked better and cost less?: Evaluating three decades of reform and change in UK central Government*. Oxford: Oxford University Press.

Isaacs, G. (2015). *The commodification, commercialisation and financialisation of low-cost housing in South Africa*. FESSUD Working Paper Series, No. 200. Retrieved from http://fessud.eu/wp-content/uploads/2015/03/FESSUD_WP200_Commodification-Commercialisation-Financialisation-Low-Cost-Housing-in-South-Africa.pdf.

John, T. (2018, May 29). How the US and Rwanda have fallen out over second-hand clothes. *BBC News*. Retrieved from https://www.bbc.co.uk/news/world-africa-44252655.

Loftus, A. (2004). Free water as commodity: The paradoxes of Durban's water service transformations. In D. McDonald & G. Ruiters (Eds.), *The age of commodity*. London: Earthscan.

Loftus, A., March, H., & Nash, F. (2016). Water infrastructure and the making of financial subjects. *Water Alternatives, 9*(2), 319–335.

Mader, P., Mertens, D. & van der Zwan, N. (Eds.). (2020). *International handbook of financialization*. London: Routledge.

MERG. (1993). Making democracy work: A framework for macroeconomic policy in South Africa. Cape Town: CDS, 1994, also New York: Oxford University Press, 1994, and translated into isiZulu as Ukwenza Intando Yeningi Isebenze, Durban: Community Law Centre.

National Housing Federation. (2017). Public expenditure on housing: The shift from capital spend to housing allowances. A European trend? Research Briefing. Retrieved from http://s3-eu-west-1.amazonaws.com/pub.housing.org.uk/public_spending_housing_europe_uk_briefing.pdf.

PAC. (2015). Economic regulation of the water sector: Oral evidence. Public Accounts Committee. HC505.

Padayachee, V., & van Niekerk, R. (2019). *'Shadows of liberation': ANC economic and social policy from African Claims (1943) to GEAR (1996)*. Johannesburg: Wits University Press.

Robertson, M. (2014). *Case study: Finance and housing provision in Britain*. FESSUD Working Paper Series, No. 51. Retrieved from http://fessud.eu/wp-content/uploads/2013/04/Case-Study_-Finance-and-Housing-Provision-in-Britain-working-paper-51.pdf.

Robertson, M. (2016). *The system of provision for housing in selected case study countries*. FESSUD Working Paper Series, No. 193. Retrieved from http://fessud.eu/wp-content/uploads/2015/03/FESSUD_WP193_The-System-of-Provision-for-Housing-in-Selected-Case-Study-Countries.pdf.

Robertson, M. (2017). (De)constructing the financialised culture of owner occupation in the UK, with the aid of the 10Cs. *New Political Economy, 22*(4), 398–409.

Ryan, F. (2019). *Crippled: Austerity and the demonization of disabled people*. London: Verso.

Serrano, A., & Brooks, A. (2019). Who is left behind in global food systems? Local farmers failed by Colombia's avocado boom. *Environment and Planning E: Nature and Space*. https://doi.org/10.1177/2514848619838195.

Unsal, E. (2017). *A political economy of electricity and housing provision in Turkey since 1980: Change, financialization and market-based social provision*. Unpublished PhD thesis, School of Oriental and African Studies, University of London.

Van Waeyenberge, E. (2018). Crisis? What crisis? A critical appraisal of World Bank housing policy in the wake of the global financial crisis. *Environment and Planning A: Economy and Space, 50*(2), 288–309.

Wilson, W., & Barton, C. (2020). Housing: Raising quality and supply. In *House of commons, insights for the new Parliament: Impartial analysis of the key issues* (pp. 14–15). London: House of Commons Library. Retrieved from https://commonslibrary.parliament.uk/wp-content/uploads/2020/01/HoP_Insights_Publication_FULL1.pdf.

CHAPTER 5

A SoP Approach to Understanding Food Consumption

5.1 Introduction

Even a casual perusal of social and mass media, let alone scholarly research and commentary from leading international organizations such as the World Health Organisation (WHO) as well as national organizations and policymakers, can lead to little doubt that we face an obesity epidemic on a global scale, part and parcel (if the most prominent element) in what has been called the diseases of affluence, eating disorders (Fine, Heasman & Wright 1996; Fine 1998). Over the past three decades, obesity has doubled, with the percentage of those considered so reaching double figures and conjectured to exceed the numbers who suffer under-nutrition on global scale.[1] Highest levels, in the Americas, well exceed a quarter of the population (and a third in the USA), with the lowest levels in South-East Asia at a few per cent. Worryingly, though, economic development would appear to bring obesity, with China's experience particularly telling with a tenth of its population now suffering diabetes, double the rate even of the UK, even though type 2 diabetes (the result primarily of being overweight, not least through consumption of sugar) was practically unknown in China two decades previously.[2] Poor diet is now the leading cause of mortality worldwide, associated with 11 million deaths, 22% of the global total, in 2017 (GBD 2017 Diet Collaborators 2019). How have we reached the state where we are eating ourselves to death? Clearly the issues are complex and go beyond the expressed preference of the individual. As Bee Wilson points out (2019, p. 6), 'it makes no sense to presume that there

K. Bayliss, B. Fine, *A Guide to the Systems of Provision Approach*,
https://doi.org/10.1007/978-3-030-54143-9_5

has been a sudden collapse in willpower across all ages and ethnic groups since the 1960s'.

For the SoP approach, diets emerge from the intersections of provisioning systems with material cultures, affecting not only what and how (much) we eat, but where, with whom and with what associated practices and meanings, let alone effects from immediate gratifications and bodily well-being to acute food poisoning or the already observed chronic diseases. As the much (mis)quoted epicure and gastronome, Brillat-Savarin, put it two centuries ago, 'tell me what you eat and I will tell you what you are', often shortened to 'you are what you eat'. The SoP approach asks how are we to explain and understand the emergence of material cultures of consumption and how do they link to the structures and processes of provisioning systems? While the specific line of enquiry for a SoP study of food will depend on the details of the research location and specific research interest, our attention here is on the effects of neoliberalism in general and of financialization in particular as one of its key characteristics. Is there any thread leading from financial markets and/or neoliberal policies to the stomach?

As always, it is possible to find contemporary problems blamed on prevailing, if shifting, grand conceptualizations—whether it be globalization or, currently, neoliberalism, with financialization looming in the wings.[3] Such structures and processes are significant for SoP analyses. This, however, raises a number of problems when addressing eating disorders: over timing given that eating disorders precede neoliberalism; over mechanisms; over what are the structures, relations, agencies and processes through which neoliberalism purportedly induces eating disorders; and over food itself (or foods themselves) for what is specific to food that makes it subject to these dysfunctions as opposed to similar or different dysfunctions for other elements of consumption (especially attached to the environment, for example).

There are two main elements in this chapter that deploys food systems to offer lessons in operationalizing the SoP approach. The first, as covered in Sect. 5.2 considers the systems by which food is provided and gives an overview of the structures, relations, processes and agents involved in food systems. This is used to argue that, under capitalism, the imperative to expand food production in pursuit of profitability inevitably gives rise to a compulsion to sell, and hence to ensure that food is consumed (and, crudely, the trend towards obesity, once beyond the stage of undernutrition), albeit with growing food waste too. In Sect. 5.2, we discuss how financialized neoliberalism intensifies these compulsions, through offering an account of the financialization of food systems, and corresponding forms of provisioning.

The second part of the chapter considers how the pressures from the provisioning system are engaged with by the consumer. The processes of expanding capitalist food production are counterbalanced by corresponding processes attached to the compulsions to control our diets with the result that food systems are marked by the tensions between the compulsions to eat *and* to diet. On one level these compulsions reflect our bodily needs to eat, but not to eat too much. However, both sides have been captured by capitalist production systems that require ever-expanding consumption but, at the same time, have developed profitable diet (and related) industries. These do not, as discussed in more detail later in the chapter, simply counterbalance one another but reside in symbiosis with one another, structuring the processes, relations and agents (including consumers) that underpin material food cultures and their attachment to the diseases of affluence. This then allows for a discussion of the nature and formation of food norms, and the role of material cultures, at the general level of compulsions to eat and to diet, and how this impinges upon some particular foods, such as sugar and dairy. A discussion of the 10Cs (see Chap. 3) around food leads to important policy conclusions concerning the weakness of dietary advice for improving dietary standards unless accompanied by strong controls over provisioning itself. The final section brings out some broader implications for operationalizing the SoP approach beyond food systems alone.

5.2 Food Production Systems: Tell Me What You (Do Not) Eat

We have previously outlined that, within a broader framework around the grand themes of globalization, neoliberalism, financialization, commodification, and so on, the SoP approach can begin to be operationalized by attending to the structures, processes, agents and relations attached to any particular SoP. Understanding food systemically has always been prominent in scholarly as well as popular discourses which have been numerous, providing more than enough material on which to (re)construct our understanding of food SoPs. This section considers the agents involved in food production before turning to processes with particular attention to financialization and neoliberalism.

In terms of the structuring of food systems, there are distinct stages in provisioning, starting with agriculture and moving through processing, transporting, wholesaling and retailing (and packaging)[4] to consumption itself (an activity that goes far beyond mere purchase). Each of these has been subject to considerable if uneven academic scrutiny, and often in isolation from one another. Thus, the 'global food system' approach, emphasizing increasingly uniform processing on a global scale in capitalist agri-food systems, especially associated with Harriet Friedmann and Philip McMichael (e.g., leading to their 1989 article) offered some inspiration for the SoP approach. But it experienced a crisis of confidence in the early 1990s because of the failure of the empirical evidence to conform to a few, relatively fixed, ideal-types of globalized commodity systems around wheat and beef for example. This, alongside the anticipated fracturing of food systems with the emergence of bio-technology, threw the whole approach into disarray. Meanwhile, the less rigid SoP approach, which specifically allowed for differences in national and product food systems, was roundly condemned by erstwhile global 'foodies' for its supposed failure to explain systemic differences by crops and country.[5] From the SoP perspective, what was striking was how different methods of provisioning, persist within and between crops even if the large-scale, capital-intensive increasingly dominate albeit alongside the small-scale, labour-intensive.

More generally, agricultural studies, whether focusing on the role of large-scale conglomerates or that of, the highly contested notion of, the peasantry, tends to confine itself to agricultural production and transformation alone albeit also addressing many themes and often their broader implications.[6] Inevitably, though, with relevance for specifying food SoPs, such analyses can reveal structuring within or around particular processes of, and relations and agents within, the food system, whether it be by access to the means of production (seed, land, fertilizer or credit), sale (to merchants) or class relations in agriculture and their corresponding structures (landlords, capitalists, landless labourers, household production, and so on).

At the other end of the food system, much emphasis has been placed upon the process of retailing and the pressures, where it is concentrated amongst a few supermarket companies, that it places on other agents, especially producers, in terms of tied contracting and low prices. Such pressures can work both ways, for example, and with supermarkets' own label (often more or less identical to branded) products, food processers are often large-scale themselves and can serve a number of customers. Who gets what out of the food system has been the preoccupation of Global Value Chain analysis, see Chap. 2, with its focusing on whether such chains are retailer- or producer-driven for example, and how rewards are structured along the chain of provision.

The state is also a key agent in food systems. In terms of health and safety of food, relations with the state are vital throughout food systems although it has been argued that large retailers are the agents increasingly taking on the role of regulation of the quality of the food they sell, not least to sustain their reputation with consumers (although this viewed benevolently took something of a hit with the horsemeat in processed food scandal in the UK in 2013).[7] Relations within the retail sector also concern wages and conditions of workers, with shelf-stacker serving as a pejorative term in reflecting how such workers are, generally correctly, perceived to be poorly paid in monotonous jobs.

While these elements have typically been dealt with in isolation, on occasion, the literature has focused on the restructuring of provision across the different structures involved, for example, to link food retail more closely with agricultural production. Clearly, large-scale retailers can play a greater or lesser role in the contracting for, and organizing of, their suppliers. In addition, there has been some emphasis upon the processes attached to the so-called industrialization of agriculture, not least through the substitution for, and appropriation of, agricultural activity by off-farm producers—seeds, fertilisers, pesticides, ripening, mechanization, and so on. Such changes are constrained by the natural, or 'organic', nature of the product and the production process especially where time is needed to allow for growth and maturing, leading to debates over how to understand relations between the social and the natural, especially in light of increasing attention to environmental considerations (how these are affected by and affect food provisioning).[8]

This whirlwind, at times implicit, and select overview of the structures, relations, agents and processes that make up food systems will have served its purpose if suggesting that, even for the same crop/product let alone across them, the ways that separate elements of food systems are put together and evolve are highly heterogeneous, across both time and place. This is so even though there may be common forces acting upon them. Rather than a food SoP, there are potentially many SoPs for different food types. However, this does not mean that general systemic aspects of the food system are inevitably overwhelmed by such complexities and varieties. Rather, systemic factors have profound causal influence, and this needs to be unpicked by SoP analyses through their contextual outcomes.

5.3 Processes: Food Production, Financialization and Neoliberalism

As with many other fields and topics, the financialization of food has benefitted from a recent and rapidly expanding literature. To some degree, both its starting point and considerable continuing weight have been placed on (third world) land grabs, reflecting how attention to this topic emerged with interest in financialization itself as well as reflecting longstanding concerns with developed country domination of developing country agriculture. Nonetheless, the literature has covered all aspects of the food chain, from land to retail (Isakson 2014; Clapp 2014; Williams 2014; Sommerville and Magnan 2015), although explicit attention to consumption has remained limited despite its being the endpoint of food chains. As Rossman and Greenfield (2006, p. 2) put it, cited in Fine (2007)[9]:

> Of course, companies have always sought to maximize profit. What is new is the drive for profit through the elimination of productive capacity and employment. Transnational food processors, for example, now invest a significantly lower proportion of their profits in expanding productive capacity. Financial markets today directly reward companies for reducing payroll through closures, restructuring and outsourcing. This reflects the way in which financialization has driven the management of non-financial companies to "act more like financial market players".

As mentioned, then, the financialization involved in land grabbing has already benefitted from its own healthy literature, Fairbairn (2015), Dixon (2014) and McMichael (2012) for example.[10] But it has already and especially healthily departed from a stereotyped understanding based on foreign financiers acquiring land for purely speculative purposes. Indeed, finance and land grabbing needs to be located more broadly across a number of dimensions. As Fairbairn (2015, p. 589) puts it:[11]

> The clear-cut distinction between foreign and domestic was always illusory ... The capriciousness of global capital makes the foreigner a moving target, difficult to pin down for more than a moment. Today, the foreigner

provides an object for popular outrage and for government policy action, but it is an unstable signifier with limited explanatory power.

In other words, the ownership of land has to be located in relation to existing as well new entrants, whether financial and/or foreign. And, for Sippel, Larder & Lawrence (2016), this is a matter of different and differentiated actors. For (p. 12):

> While some farmers appreciated the increased interest of financial actors in farmland as underpinning the value of their estates, others viewed this as a threat to family farming and an exacerbation of already unequally distributed access to land resources. Moreover, some farm family entrepreneurs have become financially attuned actors themselves and see the opportunities investment funds provide them as a means to increase their farm profitably and flexibility.

Thus, whilst some of the literature in populist vein see financialization as the latest factor undermining small-scale or native farmers,[12] Sippel, Larder, & Lawrence appropriately conclude that: 'It is especially important to avoiding constructing farmers as a singular and undifferentiated group of "victims" of a financialized food system' (p. 2).[13]

This dovetails with a rather different source of differentiation arising out of the financialization of land, one that is keenly observed in the broader literature. To what extent does financialization expand production or not (or constrain investment as for Rossman and Greenfield as typical of financialization)? On the one hand, land grabbing is seen as speculative acquisition of land that is otherwise left idle. On the other, it is associated with clearances of the Amazonian forests for the production of soya and other crops. Significantly, the farming literature to some degree offers a corrective to the notion that all financialization, especially as speculation par excellence, is at the expense of production. As Fairbairn (2014, p. 786) observes, referring to hybrids, 'it is possible to use the land productively while simultaneously speculating on financial returns from its appreciation', for Isakson (2014, p. 769), 'the financialization of farmland has blurred the line between land as a productive asset and land as an object of speculation'. She refers to the 'blurring' of the role of finance along food chains, as does Brooks (2016, p. 75) of 'financial and "real" assets and activities'. Visser, Clapp & Isakson (2015, p. 544) point out that, 'the private financial sector is conflicted about the types of risk that it is willing to assume. In particular, it is generally cautious to *invest* in actual agricultural production yet eager to *speculate* in related derivatives markets'. Sommerville and Magnan (2015, p. 135) treat 'farmland as both a productive and a financial asset', two models with different logics for them.

This raises a number of issues. What are the circumstances in which the acquisition of land by financiers leads to an expansion of production and productivity? As Li (2015, p. 567) suggests, this is a matter of '*the dynamics of change in social relations*' (emphasis in the original), and such 'relations of access and exclusion are always going to be social and political to the core'.[14] In other words, the conditions under and through which finance gains access to land is crucial, as is acknowledged by Marx's theory of agricultural rent, with the potential for financialized land accelerating rather than obstructing accumulation (Fine 1979) and see also Chap. 4 on housing.

This is a matter of how profits can best be made and, as observed in the literature, it is not simply a matter of economies of scale in farmland production but of economies of scope along value chains. Through conglomerates, financialized concentration is occurring along as well as within the separate elements of the chains of production, whether it be retailers, wholesalers, or providers of raw materials (Isakson 2014). Thus, whilst large-scale retailers have financialized their own assets (Baud and Durand 2012), they have tended to continue to rely on subcontracting for other elements in the food chain. As Brooks (2016, p. 774) comments:

> While financial actors have gained a sizeable stake in food sales ... food retailers earn more revenues from financial activities, either by redirecting cash flows into financial (rather than productive) investments or by diversifying their own activities into financial services such as credit cards, insurance products, etc.

By contrast, Salerno (2014, p. 1710) recounts another form of diversification showing how 'Cargill is a producer and trader of seed, feed, fertiliser, and agrochemicals. It is a landowner, cattle rancher, maker of transportation vehicles, biofuel producer and a provider of financial services'.

Further, food has become increasingly attached to speculative commodity markets, for particular foods and through the rapid expansion of corresponding index funds.[15] Expansion of these has been extraordinarily rapid but also shifting in nature, as particularly brought out by Lehecka (2014) with different regimes demarcated before and after 2004, and before and after the global financial crisis (GFC). What his empirical analysis shows is that (food) commodity markets can to a greater or lesser extent both track general movements in financial markets (more co-movement more recently) but also they may not track one another, suggesting commodity markets can serve as a safe haven when other markets turn down. Equally if not more important, however, is the observation of Isakson (2014, p. 759) that, in contrast to commodities such as oil, 'in most areas of the global South, the spikes in global commodity prices have translated into stubbornly high food prices'.[16] It is tempting to suggest that this is due to the way in which higher food prices can become consolidated in the form of higher land prices as capitalized rent, itself fueling financialization, and a phenomenon that has resonances with the UK housing market, see Chap. 4. And, of course, the speculative movement of the prices of different types of food may or may not be common with one another, especially in light of capacity of some products to substitute for oil.

5.3.1 *Pressure to Increase Consumption*

Under neoliberalism and accentuated by financialization, driven by the profit imperative, food systems are subject to increasing expansion of supply and cost reduction, for given quality (itself subject to branded invention). But, with the exception of the occasional, deliberate destruction of yield to sustain prices or heavy recession and stockpiling, food that is produced needs to be consumed. As a result, commodified food systems are driven by the compulsion to sell and, thereby, providers have an incentive, if not need, to create a social compulsion to eat that goes far beyond physiological need. The question is not so much whether this is so but how it is realized socially and systemically. Through what processes are food relations structured so that consumers are agents that systematically eat more and more? Three observations are made here.

First, as shown in a study of food systems through the SoP approach in the mid-1990s, if fat, sugar and salt are produced (e.g., in dairy, sugar and meat systems), then they are going to be consumed. It is simply a matter of by whom and how, especially in processed foods, for example, where they become cheaper, disguised and make and serve palatability and convenience. For the latter, the rise in the UK of supermarket retailing had ensured the availability of a wider range of healthier low-fat milks, once the uniquely protected doorstep delivery was effectively abandoned. But with the agricultural system supporting production of high fat milk, the cream had to go somewhere. And so it did, into fancy cheeses and desserts, and manufactured foods, all equally readily available in the multi-product supermarkets. Indeed, those at the forefront in the purchase of healthy milk were found to be equally prominent in the consumption of high cream products as well (Fine, Heasman & Wright 1996)! A similar story can be told for sugar, with reduction in direct consumption from the sugar bowl or in home-baking being compensated for by its incorporation within manufactured foods (alongside salt and unhealthy fats), sustaining its level of consumption per capita. Thus, analysis of the dairy and sugar systems indicated that healthy eating programmes for the consumer would tend at most to redistribute their consumption, and most likely towards those on low incomes, with poor diets in the first place, and least able or willing to respond to health messages.[17] This provided one major insight into food systems, as the so-called diseases of affluence associated with bad diets came to be recognized, namely the inconsistency between agricultural policy (produce more as cheaply as possible) and health policy (eat less and a balanced diet) with agricultural production (and food processing) prevailing over health.

Second, rising food production is associated with increasing availabilities of food. It is not only the different ranges of foods available, throughout the year without regard to seasons, but the appropriately dubbed convenience with which they are made available—through preservation as dried foods or in cans, by storage in fridge/freezers, ready meals, snacks, and ever more outlets for the provisioning of foods, by small and large stores (one for convenience, the other for bulk purchase),[18] with the Deliveroo/Uber phenomenon being the most recent manifestation of the capacity to eat whatever you like whenever you like, all on top of take-aways, cafés, street foods, quite apart from home cooking in increasingly well-equipped kitchens with ever more ingredients and recipes to deploy. Third, the food industry is also compelling us to be healthy, not necessarily by consuming less but via the expanding range of diet foods on the market.

5.3.2 *Compulsions to Eat and Diet*

Financialized neoliberal food systems are creating pressures to increase food consumption. However, given the confines of appetite and bodily functions (and corresponding health), the compulsion to eat has become complemented by an equally pervasive compulsion not to eat, but to diet or, at least, to eat healthily. There is a sense in which the food system is seeking to turn us all into food addicts. Significantly, for most legal or illegal 'addictions' (drugs, alcohol to a lesser extent, smoking, gambling), the majority of profit tends to be made from the minority of consumers, and these can be targeted by suppliers as a way of reconciling the general consumers' compulsions both to indulge and to refrain. In the case of food, we are all addicts and non-addicts at the same time—we cannot go cold turkey as we have to eat but we also have to direct if not contain the fulfilment of our indulgences through food and eating. The issue is how we receive, experience and respond to the dual compulsions to eat and to diet, and the circumstances in which we do so, and through the consumption of which foods and how. In a nutshell, how are we to understand the material culture of food given the symbiosis between the compulsions to eat and diet and how are we to explain them?

Whatever the historical origins of the compulsion to diet in concerns for health and well-being, the twentieth century food system is characterized by precisely the same intensified processes of commodification characteristic of financialized neoliberalism. In this respect, the food and diet industries are not only in some sense in conflict with one another but are also symbiotic and mutually supportive. And this is an essential starting point for understanding social norms of food consumption. How the tensions between the two compulsions are resolved is vital in understanding the position of individual foods in our diets even if, as is apparent from obesity statistics (the very construction of which is itself evidence of the tensions surrounding our diets!), the compulsion to eat is of greater quantitative significance (just as climate change can lead to extreme weather events, including those attached to low temperatures but attention tends to focus on global warming).

5.4 From Financialization to Food Cultures

The financialization of food can be seen both to share the leading characteristics of financialization in general and to exhibit peculiar characteristics of its own, related to social relations on and from the land and the nature of agricultural production and food processing themselves. Without going into sectoral detail, this all serves to reinforce the conclusion of highly differentiated outcomes across product, place and time and along corresponding value chains. However, the financialization of food literature does not even begin to engage with the material cultures of the consumption of food.[19] Rather, the financialization of food literature is, in traditional terms, primarily concerned, if admittedly in commendable detail, with the conditions of supply and demand only up to but not beyond the point of sale. And if wandering, if not wondering, into the dark territory of consumption itself, the focus tends to be on too high prices and too low incomes to be able to avoid hunger, or the satisfaction of ever-expanding elite tastes for all-year round availability of seasonal or niche products, ideally sustainably, organically or ethically delivered whatever this means in terms of increasingly commercialized and contested qualities of the foods and their provisioning, and the conditions of workers and farmers.[20]

While eating disorders (over and under eating) do not originate with financialized neoliberalism, they are intensified by it through literally feeding, both materially and culturally (see below quite apart from own common experience), the compulsion to eat. But what of the simultaneous pressures not to eat, to diet, or whatever, that at least have some bounds in bodily capacities and concern for own well-being beyond nutritional satisfaction and indulgences?

Here, it can be argued that financialization has also intensified the compulsion to diet, not least by reference to the commercialization of food. Long before neoliberalism, even with (capitalistic) commodification of food systems on global and national scales, food provisioning and practices have substantially exceeded the buying and selling of food for the purposes of consumption. For example, commodification has brought the ready-availability of food ingredients for preparation of foods within the home, alongside an ever-expanding paraphernalia of devices and gadgets, from ovens to cookery books, to allow for domestic preparation and serving of food. As a result, paradoxically, the commercialization of the food system has equally underpinned non-commercial provisioning of food, or at least, certain aspects of final preparation of readily available ingredients even to the point of ready-mix, ready-to-eat and so on. Thus, the capacity both to eat at home and to prepare food at home has been extended, not least through home provisioning with more or less ready foods and the means with which to prepare them for consumption being made commercially available.

From this, two important implications follow. On the one hand, there is the expansion of non-commercial forms of provision, as a result of which there are ever-more forms of availability of food, buy it or make it, intensifying the availabilities of foods (although such forms of provisioning tend to compete with, and be displaced by commodification). On the other hand, broader ways of provisioning food equally allow for the targeting of healthier diets, intensifying the pressure to avoid unhealthy foods whether (partially) self-prepared or not. This also sits alongside domestic commercial or economic calculation upon how we access our food, not least in getting value for money quite apart from any qualitative considerations around the merits of commercial as opposed to home-made food. Indeed, the forms taken by the compulsion to eat are so extensive that they even extend to what might be termed the negation of commercial considerations altogether, that we eat without regard to cost as with celebratory meals whether in a, paradoxically costly, restaurant meal or weddings at enormous expense, or at home where what is provided similarly takes absolute priority, as with Xmas dinner, and the like, where there are tensions between budgeting and avoiding the stigma of monetary considerations.[21]

In light of the above, it is tempting to lay the blame for eating disorders at the door of financialized neoliberalism, as producers compel us to consume ever more quantities of food produced, in whatever innovative forms. But the reality is more complex. Indeed eating disorders predate neoliberalism, by decades in both incidence and diagnosis, although, as stated the past three decades have been associated with a rapid expansion in overeating. For SoP, therefore, attention is needed to the ways that we engage with the production system. And this raises issues over how to approach the impact of neoliberalism on the material cultures of food in general and eating disorders in particular, bearing in mind that food exists over a huge range of cultural practices and beliefs, including the rise and fall of fashions alongside secular changes (in eating out, e.g., convenience foods, takeaways and deliveries, the displacement of work canteens, the rise of the snack and the sandwich, the coffee shop, fair trading, the organic, the celebrity chefs and diet gurus, food scandals including mad cow disease and horsemeat, and so on).

Of course, food is not unique in exhibiting longstanding dysfunctions that predate neoliberalism whether it be the various aspects of poverty and deprivation, climate change and fossil fuel capitalism, unemployment, and so on, although there is a tendency, as already observed, to blame neoliberalism for these just as previously globalization and even Keynesianism might have been blamed. And there have been attempts to associate eating disorders with neoliberalism, as with Guthman and DuPuis (2006) in which, in Foucauldian fashion, the individual is seen literally to embody the imperatives of (US) capitalism.[22] But such an approach, quite apart from needing to specify neoliberalism itself more precisely,[23] begs more questions than it answers—not least why do eating disorders predate neoliberalism, do they prevail over all sections of the population, are things different across place and time (is the USA the exception or the prospect for elsewhere), what about countervailing resistances, measures and policies, do such considerations apply to all or only to some foods and why, when and how?

And this is an essential starting point for understanding social norms of food consumption. How the tensions between the two compulsions, to eat and to diet, are resolved is vital in understanding the position of individual foods in our diets even if, as is apparent from obesity statistics, the compulsion to eat is of greater quantitative significance. Food norms can now be seen to be highly complex. As emphasized more generally for consumption norms for all goods, they vary by who gets what and how. But, further, in light of our analysis, especially for food, the very nature of the norms is product-specific, in examining how consumption relates to the underlying compulsions to eat and to diet. This is apparent in case of foods like sugar and dairy but applies in different ways, for example, for the tea, coffee or hot drink SoPs. Are they taken to avoid calories, with or without sugar and/or biscuit, a smoke, for sociability, to ease boredom? In other words, this places us in the world of the material culture of food systems and practices to which we now turn.

5.5 Material Cultures of Foods

5.5.1 10Cs and Food Consumption Cultures

The most likely best starting point for a material culture of consumption in general, and of food in particular, in light of the 10Cs, is that consumption cultures are Constructed and in a dual sense. On the one hand, as we have already heavily emphasized, contemporary food systems are based upon the Contradictory pressures to eat and to diet. On the other hand, this is equally heavily important in conditioning how we form our food beliefs and practices. The construction of the MC for food comprises three elements—how food is materially provisioned, how it is attached to practices and meanings, and the relationship between the two. As argued, food systems push for the expansion of profitable production against what are bodily capacities that have to be yielding to excessive consumption to the point of what has become an epidemic of obesity in order to accommodate profitable supply. As a result, the availability of food has become pervasive, in terms of what can be eaten, how and when. But neoliberal or other governance, regulation or advice to the contrary, such outcomes through consumption themselves have to be reproduced or, possibly, transformed. Why do food cultures remain the same, or change, in light of how they are constructed? What are the tensions involved in negotiating the compulsions to eat and diet? This is the light in which to situate obesity, and eating disorders more generally, through social epidemiology or the social determinants of health.

However, initially observe three things we are not trying to do. One is that we are not seeking to address individual susceptibility as such to obesity as opposed to the structures, relations, agencies and processes through which obesity is generated. For the individual, as is only too commonly observed, that is more or less the work of the personal psychiatrist as opposed to the social psychologist, the dietician, the personal trainer, the doctor or surgeon, fashionista, family or friends, and own knowledge and practices (although all of these are liable to be that much weaker or even counterproductive for not acknowledging the imperatives of food systems and their material cultures). Second, nor are we seeking to identify the likely foods that are associated with poor dietary outcomes, nor the likely locations or incidence across social strata. However, anorexia nervosa and bulimia originally were primarily focused on girls entering womanhood although these have spread unevenly across more or less all sections of the population. Third, nor do we more than outline the processes through which food is delivered and sold to us, across weekly to corner shops, and snack to family meal although we observe this is, and has been, the work of the SoP approach to different foods such as meat, sugar and dairy (Fine, Heasman & Wright 1996).

More constructively, what we can do, is to view eating disorders as symptomatic of the construction of food systems as a whole, not least given the ubiquitous presence of the compulsions both to eat and to diet. Whilst some lines (and this is diagnostically difficult, vague and controversial) may be drawn between those who suffer an eating disorder and those who do not, the important point is that eating disorders—both over- and under-eating—could in part be considered as the more extreme, individual forms taken by food systems, rather than a pathology that distinguishes the good from the bad, the ill from the healthy, and so on. By way of analogy, pollution is a modern phenomenon that is a social and material determinant of breathing disorders, as is obvious irrespective of which individuals succumb or not, with some worse hit than others in part because of life choices and psychological dispositions (urban environment, smoking, etc). For the SoP approach, the focus is at the *systemic* rather than the individual level, and parallels could be drawn with the Covid 19 pandemic (see Chap. 6) as well as the social determinants of health more generally.

It is not just the ubiquity of food that is characteristic of modern food systems but its corresponding capacity to serve as a ubiquitous focus for all of our emotions, more or less without exception. It starts with mother's milk (or not with the deplored and deplorable promotion of formula to displace it) and has an unavoidable presence through the feasts at all of life's major events from birth to death, as well as ever-expanding traditional or newly adopted blow-outs in the commercial patterns set by the US celebrations of mother's and father's day, Halloween, Christmas, New Year and whatever else can be incorporated as an ever-encroaching excuse for excesses for food consumption and otherwise, from dating to doing something special, or as a more or less regular treat. Traditional overconsumption at more and longer festivals has its counterpart in the regular patterns of daily life, not only through the mechanisms for provision already laid out (cheap food, anytime and anywhere) but as the way increasingly through which to focus upon our emotional, and not just our physical well-being, with whatever degree of success or, ever more often, failure as indicated by the consequences in terms of both obesity and the attempts to avoid or to respond to it.[24]

And what is true of the generation and satisfaction of our emotions and the emotional extends to the cultural more generally as with the family, the national, the ethnic, comfort or mundane eating, and so on. In short, food has become what might be termed a key element in the material and cultural processes for gratification of the emotions as well as, if not more than, for gratification of the stomach. In this sense, food has become both pliable and extensive in the cultures it can serve, precisely as both cause and effect of its ubiquity. By the same token, the material culture of food is subject to be Construed. Foods do not arrive with given meanings as such, although how they are provisioned conditions what meanings they can have. After all, we shudder, or joke, at the idea of road kill as a source of our meat but suggestions of farm kill are studiously avoided by the neat packaging of meats to appear as more or less inanimate for most sales (apart from images of healthy, living animals on the packaging, and, as such, at least one valid example of the omnivore's paradox—we do not know of the bloody and other details of where and how our meat comes, labelling aside, and we generally do not want to know).

Nonetheless, vegetarianism in its different forms is alive and well, indicating that consumers are not passive recipients of the meanings of their foods. Even so, in light of the cliché referenced earlier, 'tell me what you eat and I will tell you what you are', the role of construal in constructing the meaning of foods necessarily complements, in part forms, the cultures and identities to which the consumer is attached. Yet, because Commodified food has to be bought and so made palatable, advertising and promotion of its properties in any numbers of ways are pervasive (not least through the eating and cooking industries and its popular projection through all forms of media). As discussed, commodification is complemented by commodity forms and commodity calculation, underpinning the ubiquity of food, with Conformity to commodity logics even where they are not present or even rejected (home-made for shop bought). Even more important is how counter-movements tend to become incorporated into commodification as with fair trade, environmental standards, and so on, each of which becomes subject to redefinition, if not dilution, to support continuing profitability.[25]

Inevitably, given contradictory compulsions to eat and to diet, how these are culturally embodied is Contested although in ways that are Chaotically incorporated into the meanings assigned to foods. The diet industry plays a role in this with its corresponding, and equally obsessive attention to body images with spin-offs ranging from the exercise industry to fat-removing surgeries, with body image and health themselves a source of construal and scrutiny. Further, however, it is defined, the search for ethical consumption is riddled with incoherencies around equity, fairness, human (and animal) rights and so on, such that to be ethical in principle would even prejudice mere survival in practice (see Chap. 4 for discussion on ethical consumption in relation to clothing). This is not just a matter of compromise but one of reconciling competing demands upon our cultures, something that cannot be resolved merely through consistency and clear thinking. Indeed, such would appear to lead to the hermit or the purely hedonistic however (socially) defined and pursued![26] And, correspondingly, while cultural systems are chaotic in the sense of being riddled with compromises, inconsistencies even, of beliefs and actions they are not, thereby, rendered arbitrary nor open, given how heavily they are conditioned.

This chaos is individually experienced but it is equally a set of Collective angsts surrounding the consumption of food. Cultural influences, as already observed, are generated at a distance from consumers, and trickle-down accordingly as opposed to only being actively initiated and made by the independent consumer. This is obvious in the sense that we do not, for example, construct meanings by advertising to ourselves even if we do ourselves interpret advertisements. By the same token, ways of constructing meanings, beliefs and actions tend to preclude but not necessarily to prohibit other ways. Just think of trade-marking, standards, branding, regulations, and so on, not just what they are but how they are made and who gets to make them. In these, and other respects in terms of the material nature of commodities, it is as much a matter of what is not revealed—and by whom, how, and why—as it is revealed. Buy an 'orange' drink with the proclaimed selling point of 'natural ingredients only' and the chances are it contains no oranges and is full of sugar despite the attempt to appeal to the health moniker 'natural'.[27] Indeed Foster, McMeekin & Mylan (2012) in their SoP analysis of orange juice discover that attempts to encourage more sustainable patterns of production and consumption by 'carbon labelling' had little effect due to path-dependent consumer expectations and their links to 'wider food-related practices' (p. 402).

These considerations range widely over the different Contexts through which foods are provisioned, from treat through family meal to breakfast on the run.[28] Inevitably, the attempts to influence outcomes are collective, from those who promote profit as against those who promote healthy eating, these serving not so much to increase choice and opportunity as to deepen the stresses around food. And those stresses are confronted in ways that are deeply conditioned by Closed access to influence, with the budgets and reach of the food industries far exceeding those of governments whether in advice or policy not least in ideologically dubbing corresponding interventions as those of the nanny state at the expense of our freedom and enjoyment.

As argued, financialized neoliberalism has intensified, rather than originated, these aspects of food systems as can be seen by examining the pathways through which eating disorders are socially and individually generated. First and foremost, as is characteristic of financialization more generally, the pursuit of short-term profitability to underpin speculative potential inevitably leads to the intensified promotion of more and cheaper foods (and pressures on those who work along the food chains) without regard to the consequences for health. This gives rise not only to overeating (and pressures to diet) but also to the presence of undue quantities of fat, sugar and salt in what we eat. This is not necessarily a deliberate conspiracy to make us unhealthy, although there are shades of this, for modern food is flush with both water (for added weight) and air (for added volume) with no direct ill consequences other than regulated deception (what is meat when 20% water, or bread if similar for air).[29]

Second, though, these imperatives inevitably impinge both directly and indirectly upon our food habits. This is so even where profit is not immediately present, as own preparation of food is subject to commodity form and calculation, not least as so much of our food consumption depends at least partially on our own contribution to provisioning—we buy and prepare with ready-made meals or ingredients with more or less care and attention to their nutritional content and relative and absolute costs. Perversely, just as food is increasingly projected as a neoliberalized individual responsibility, so it becomes a realm of expanded choice deeply embedded in closer or more distant consequences of the pursuit of financialized forms of profitmaking and their impacts on consumption norms.[30] For example, eating patterns in the UK have become increasingly chaotic, in part due to a huge increase in unstructured food consumption through snacking. This is related to greater availability of food and changing sale patterns with, for example, the rise of the 'multipack' (Pirie 2016).

Third, as observed, there are equally contradictory pressures to diet, at individual and social levels, leading to reactions and stresses that stretch from the social to the individual, especially in how food is regulated, from labelling through portion sizes to taxation. Taxes on sugar in drinks, for example, will shift its presence into other ready-prepared foods, quite apart from the increasing role played by corporate interests in the processes of both creating our cultures and beliefs and in policymaking. Each of these has involved a double shift under neoliberalism in the form of stronger corporate role and influence as well as their attachment to financialized forms of accumulation. For the tip of the tip of the iceberg of such influence, consider that numerous studies have found that sugar-sweetened beverages contribute to the development of obesity and type 2 diabetes (see MacGill 2018 and citations therein). Some studies however, such as the meta-analysis by Rippe and Angelopoulos (2016) find, controversially, no such association but then disclose that their work has been funded by Coca Cola among others.

Stevano, Johnston & Codjoe (2020) also highlight the power of the food industry when it comes to narratives regarding nutrition. They research the drivers of eating habits of school children in Ghana. They find that there is an aspiration for 'continental' food (such as pizza and KFC) over 'local' food among children, with the former more likely to be consumed by wealthier children. Furthermore, food companies had a presence in nutrition narratives, with, for example, Nestlé sponsoring a 'Healthy Kids Programme'. They conclude (p. 17):

> the intersection between rising inequality and the unscrutinised concentration of power in the hands of the food industry contributes to polarizing material conditions of daily life and to forging cultural relations to food. Furthermore, the material constraints, the concentration of capital and power, and the manipulation of discourse strongly suggest that conceptualizations of human behaviour as a sequence of individual decisions is rather unconvincing.

5.5.2 *How (Not) to Promote Healthy Eating Habits?*

In light of the above, how should eating habits be improved? The conventional view, underpinning most policy and buttressed by ideas of freedom both to choose (what to eat, as motivated by notions of consumer sovereignty) and from the nanny state (telling or forcing us what to eat),[31] is that healthy eating campaigns can affect positively food beliefs and behaviour in particular.[32] It is argued that healthy foods would drive out the unhealthy, just as correspondingly truth-based, possibly regulated, advertising would drive out the false claims underpinning unhealthy choices (with children's breakfast cereals packed with sugar to the fore, for example). Yet, we certainly know this has not occurred in practice. Will more information improve our diets?

Fischler's (1980, 1988) so-called 'omnivore's paradox' suggests that our knowledge of nutrition has been eroded by our distancing from food production processes. While we can, as omnivores, eat more or less anything, we need to be cautious to avoid poisoning contingent upon our knowledge of food. But he argues, our knowledge, and its attachment to potential harm or distaste (think horsemeat in our burgers), has been increasingly undermined by our detachment as consumers from food production, itself a consequence of commodification in general and globalization in particular, bringing physical as well as knowledge distance from the origins of our diets in the processes of production and distribution.

However, as Fine (1993, 1998) argues, by reference to what he terms the 'diet paradox', our knowledge of the food we consume, insofar as these things can be quantified, is of a considerably greater magnitude in many respects than that of our directly producing and consuming peasant ancestors if such they were. For many, our current common knowledge of nutritional properties of foods is vastly superior to that of olden times however nostalgically viewed. This is because direct knowledge and experience of production (and processing, wholesaling and retailing) are not the only sources of knowledge in general and of these elements in particular.[33] Although *how* our knowledge, as one aspect of the material cultures attached to food, is generated remains to be explored (as descriptors like fair trade, organic, natural, etc, and healthy itself are themselves endowed with meanings as much by factors other than the physical properties of the foods themselves and how they are provided).

Thus, while we may have a healthy diet that we target with greater or lesser success, in practice, clearly we do not organize our eating habits around whatever knowledge we have. Such is the rationale for posing a diet paradox in place of the omnivore's paradox since, whilst everyone does have a diet ex post, it is not necessarily determined as such by ex ante intentions. It is arguable that people know very well what constitutes a healthy diet—the problem is that they simply do not observe it.

Given the range of factors that shape food consumption, it does not make sense to understand the determinants of diet by more or less deviation from what is projected to be ideal or healthy. Food beliefs do not derive exclusively, nor even primarily, from individuals struggling with their knowledge of their putative deviations from a targeted, ideal diet. Most people know whether they are eating healthily or not, and most also know whether they are going to try and do something about it. A targeted healthy diet might be good for the individual consumer but not, in general for understanding what is eaten. In general, what we eat is not determined by the greater or lesser success in targeting an ideal diet, so there is relatively little analytical purchase in trying to understand what we eat by deviation from we ought to.

By contrast to this ideal-type understanding of diet, dietary knowledge derives from a multiplicity of sources, and corresponding practices and experiences (including those attached to foods not nutritious to eat or drink). While, food knowledge is one thing, eating patterns in practice reflect the range of material cultures of food systems outlined above. But how, what and why do these come about?

In early food SoP work, a starting point was made, in deference to interest in the potential impact of healthy eating campaigns, with the idea of an 'information system' attached to each food SoP (Fine 1998). In the context of broader sources of information, not least corporate and brand advertising, for example, diagnosis of such food information systems seeks to locate the favourable or unfavourable promotion of foods or food ingredients. So too much meat might be considered unhealthy, or some ingredients of foods are looked upon unfavourably, and others favourably within limits. Indeed, this is all indicative of the diet paradox. We are told to eat less salt. But while people do not eat salt and decreasingly sugar and fat, and so on directly, they increasingly figure as ingredients in processed, convenience or ready-to-eat foods, thereby boosting individual consumption. And this is important in terms of both beliefs, actions and the evolving relationship between the two.

It does not take much, however, to realize that reducing food beliefs to information systems is at the very least terminologically inadequate. Food information systems (reduced to dietary advice or not) are a poor proxy for the much more widely cast notion of the material cultures of food, in which information as such is not even central. Interestingly, the push within the SoP approach to remedy the divorce between (the intentions of) food information systems and the material cultures of food arose initially out of the attempt to explore the gendering of (food) consumption. This revealed just how much the weight of academic literature had been focused on gendered consumption around alcoholism, smoking, and eating disorders, much more than the clothing fashion system for example. Eating disorders in particular have generated significant sociological and psychological literatures, proving to be empirically informative around both evolving incidence and proximate causes (e.g., why young women disproportionately suffer eating disorders). However, these fail to confront the decisive role played by food systems in their modern form. In particular, a political economy of eating disorders was proposed in response in which, it was argued, all individuals, in advanced capitalism at least, need to negotiate the simultaneous, unavoidable and increasingly powerful tensions both to diet and to eat, a decisive factor being the capacity of food systems to promote both of these commercially albeit not necessarily at one another's expense (Fine 1995, 1998).[34]

Of course, this is not to reduce food beliefs, in the context of eating disorders or more generally, to these contradictory tensions, only to set a context for other factors. The tensions to eat and to diet do not derive exclusively from food systems as such, individually and collectively, but each food SoP has a very different relationship to these tensions that shifts over time. In short, food systems underpin consumer cultures through which pressures both to consume and to abstain need to be (re)negotiated. Thus, our associations with foods are shaped by the SoPs by which they are provided and the reality of food consumption is recognizably a different world of food cultures than one of deviations from an ideal and satisfying diet.

This returns us to the theme with which we began this chapter, eating disorders which have become endemic with the numbers of those, globally, now suffering malnutrition through overeating matching those subject to malnutrition. Significantly, the nature and incidence of eating disorders, narrowly conceived, and whether they are characterized as such, shifts across time and place. Essentially, though, it is no coincidence that eating disorders are a peculiar phenomenon of the twentieth century or, at the very least, only take on the form, content and incidence that they do, under conditions of advanced capitalism.

5.6 Conclusion

As discussed in this chapter, at an early stage in the development of the SoP approach, it engaged in the 1990s with the preoccupation with (improvement of) the Nation's Diet inevitably placing the culture of consumption back on the agenda.[35] But, in the first instance, concern focused on the relatively narrow terrain of what impact could healthy eating campaigns have on what were rapidly becoming or, had already become, the dietary diseases of affluence associated primarily with overeating and poor diet (excessive salt, fat and sugar, etc) with, for example, onset of middle-age diabetes, high blood pressure, cancers, and so on, more or less unavoidably disastrous from social and individual perspectives.[36] Most campaigns for promoting healthier consumption sought to close the gap between the ideal and actual diets through promoting the former and, possibly, the deleterious effects of the latter (addressing the diet as opposed to the omnivore's paradox). And these have failed miserably in part for systemic reasons outlined previously—what gets produced must more or less get consumed, and can be more a matter of expanding and redistributing the burden of fat, sugar and salt consumption, underpinning the diseases of affluence as marked by the contemporary global epidemic of obesity.

Corresponding problems have intensified over the last two decades. Modern food systems have led to diseases of affluence, which have underpinned norms of consumption, with the ubiquity of food generated through commodification, and intensified as such through financialization. For it is the contradictory pressures to eat and to diet that marks the twentieth, and twenty-first, century diseases of affluence. The processes through which the compulsion to eat materializes are inescapable aspects of our daily lives ranging over the ever-expanding availability of foods through the day as a result of convenience (foods themselves and ready-prepared for micro-waving or otherwise), snacks, take-aways, doorstep delivery of meals, and so on. This is complemented by a diet industry that often involves eating something alongside rather than instead of your 'normal' diet.

Indeed, at times, the contradictory and chaotic nature (and attachments to the other eight of the 10 Cs) of the cultural consequences of food systems are sharply revealed. Thus, is it accidental that the two most popular forms of non-fiction are cooking and dieting books; that we are inundated with encouragements both to eat and to diet; that eating and dieting are made compatible with one another by making both a healthy meal choice and adding on an indulgence as a reward for doing so (burger plus diet coke); that a snack is not a meal, and so on. At other times, cultural determinants are less obvious as with the influence of store wars (see Chap. 1) on the forms and levels of food advertising.

The SoP approach has much to offer in specifying and studying food systems, both in general and for particular products or product groups. How the approach is used, and for what, depends upon the research question, or possibly policy initiative, at hand, with implications for where the boundaries of the food SoP are drawn. It makes sense to take dairy as a whole but not sweetness, as sugar belongs to an entirely different SoP than artificial sweeteners. And the SoP approach may be useful for, but has limitations around, the analysis of eating out (and delivered in), given how shifting, differentiated and fragmented are its constituent components, although this too is best addressed through the practices, cultures and their interactions of provisioning/accessing (Warde 2016).

In addition, the SoP approach to food can be seen to belong to a more general category of what might be termed the political economy of excess, if again not originating with, but being intensified by, financialized neoliberalism. We can see that the SoPs for energy and clothing (see Chaps. 4 and 6), and most notable of all, for finance, are subject to excesses but, equally, with parallel compulsions to contain. Contrary to the broad thrust of the financialization literature, and its presumption other than for finance, that financialization has a dampening effect, the opposite seems to be the case. Yet, carbon trading, for example, is about allowing for expansion of both clean and dirty energy and, by the way, consumers pursuing clean energy or energy products creates the space for more dirty energy in production.[37] For clothing, fashion in contemporary capitalism is always caught between the demands for novelty, differentiation, emulation and cheap mass production as a route to greater profitability, just as we want both throwaway clothes and those of quality that last (Fine and Leopold 1993). And the material culture of finance is fundamentally based upon the competing pressures to save and yet to spend upon credit (Bayliss, Fine & Robertson (Eds.) 2018).

Notes

1. For the evolving understanding of the incidence and (capacity for) treatment of diabetes, see the regular and increasing numbers of contributions to the *Lancet* and, especially and significantly, meta-studies. Note that it has begun to emerge that the issue might be related to the food system if without further analysis, itself reflecting a reactive, curative approach by health care to the social determinants of health, see Chap. 4.
2. For an account, sympathetic to the SoP approach, on how 'development' brings 'bad', commercialized food habits to the South, see Stevano, Johnston & Codjoe (2020) and Stevano (2020).
3. See Guthman and DuPuis (2006).
4. Apart from the highly contested labelling, packaging tends to be ignored, despite its ubiquitous materiality, unless a problem (increasingly with environmental damage through plastic), see Murcott (2019, p. 98).
5. See debate in special issue of *Review of International Political Economy*, 1 (3), 1994, pp. 519–86.
6. See, for example, contributions to journals such as the *Journal of Peasant Studies* and the *Journal of Agrarian Change*.
7. See Marsden, Lee, Flynn & Thankappan (2010).

On the industrialization (and globalization) of food systems, see Cramer and Sender (2019), Cramer, Johnston, Muller, Oya & Sender (2020), Barrett, Reardon, Swinnen & Zilberman (2019) and Wrigley and Wood (2018). Note that the industrialization of food approach gave way in the work of Goodman to an assault by him and others on the SoP approach, inspired by Actor-Network Theory (ANT) focusing on the supposedly invalid separation of nature and society in the approach and more generally (Goodman 1999, 2001, 2002) and Goodman and Dupuis (2002) and see

also especially the special issue of *Sociologia Ruralis*, 42 (4), 2002). See Fine (2002, 2004) for a defence of the SoP approach against these charges, and, for critique of ANT more generally, Fine (2003a, b, 2005). With tongue in cheek, it is apparent that the ANT SoP has evolved into the performativity approach, with its leading focus being upon finance, and the idea that economists make (financial) markets, having shifted focus away from food in general and the performing scallop in particular, from Callon (1986) to Callon and Caliskan (2005), the latter incomprehensibly claiming that, 'What happens in the sector itself, how its markets are organized, its prices are set, networks built, research carried out are still left untouched in the systems of provision approach'. See Fine (2016) for a critique of performativity in passing, and Brisset (2019) for a balanced assessment. For the severest criticism of the policing of the nature/society dualism, see Malm (2019) and Fine (2019) for similar issues arising out of criticisms of Marx's apparent neglect of the natural for the social.
8. See also Newman (2009).
9. On land grabbing more generally, see also collection introduced by Edelman, Oya & Saturnino (2013).
10. See do Nascimento, Frederico & Saweljew (2019) for the Brazilian case but also IDI (2017) for the indirect support offered to such land grabbing through financing from the World Bank.
11. As Sommerville and Magnan (2015, p. 137) put it, 'financialization accelerates the concentration of land and resources into fewer and larger farming units'.
12. This differentiation across and within the food systems in promoting and in response to financialization is explored in some detail in Baines (2017).
13. With the role of the state crucial in this respect, as Visser, Clapp & Isakson (2015, p. 544) polemically observe, 'states' objectives have shifted from ensuring that agriculture is supported by finance to ensuring that finance is supported by agriculture'. Thus, subsidies to production can be capitalized in the value of land and securitized as a financial asset on condition of expanding, or constrained, production.
14. Isakson (2014, p. 758).
15. See also Ghosh (2010). However, the trend in food prices, necessarily roughly estimated, has been downwards, by about a third in real terms, over the past fifty years prior to the spikes around the global financial crisis. Burgeoning rent and land prices come with productivity increases at rates higher than price decreases.
16. See Fine, Heasman & Wright (1996, 1998) and Fine (1998) for these and other studies and their many implications. Guthman (2015, p. 2522) quotes from a US comedy show: 'The government is really sending mixed messages here. First, they subsidize corn, making it so cheap we can gorge

on subsidized corn syrup, and then they charge us more for health insurance just because our organs have caramelized ...Well, I'm sorry', he quipped, 'but our bodies are the only growth industry America has left'.

17. See discussion of 'store wars' in Chap. 1 and notice how, in the UK context, convenience stores have increasingly become branches of the main stores, in petrol stations or standalone in urban settings with longer opening hours.
18. It does, however, engage with the material cultures of food or, more exactly, its production, and so on, short of consumption. Thus, often in the context of distancing, see above, (financialization of) food has become set in cultures of (in)security and the fear of the foreign (land grabs) (Larder, Sippel & Lawrence 2015), and to alleviate poverty and insecurity and promote food sovereignty through the private sector (Brooks 2016), with the World Bank to the fore with 'a key role in advocating for more commodity exchanges in developing countries at the same time that it has portrayed large-scale land acquisitions as an important development opportunity' (Clapp 2014, p. 806).
19. Carolan (2020), focusing on the links between food production and food consumption and noting that Walmart is the largest provider of ethical foods in the USA, finds very different and contested notions of ethical (or fair trade) foods between consumers and farmers. Indeed, although using a moral economy framing, he might just as well be describing the SoP approach, p. 18:

    > economic institutions and practices are founded on norms that instil within markets understandings of 'the good' and 'the right,' generating effects that have implications that touch on understandings of self-worth, status, subject-hood for actors throughout the value chain.

    On conundrums around the material cultures of ethical foods, see Cramer, Johnston, Muller, Oya & Sender (2017) for limited impact on wages and working conditions and also Soper (2019, p. 328) for whom:

    > Conventional strawberries are larger and therefore fewer of them fill up a box. Farmworkers routinely pick more boxes in conventional than in organic, thus earning more, since under the piece rate system, farmworkers are paid per box. With short-term economic survival rather than long-term occupational health concerns in mind, strawberry harvesters would rather work on conventional farms because "la fresa orgánica es más chiquita" (organic strawberries are smaller).

20. Note that modern food systems expand both commodification (commercial provision both extensively and intensively) and commodity calculation

(more capacity for own provision) but do not tend to underpin more casual commodity forms in food provisioning (charging neighbours for food, for example, or second-hand markets for food).

21. A more rounded approach is taken in the later Guthman (2015) if drawing upon Harvey in suggesting the body as a socio-ecological fix for agrofood capitalism. See also Guthman (2011, p. 182) for which, 'The body is part of the spatial fix', and Fine (2006) for a critique of the proliferation of fixes deriving from Harvey's original spatial fix.
22. For our own take on all things neoliberal, see Fine and Saad-Filho (2016), Boffo, Fine & Saad-Filho (2018) and Bayliss, Fine, Robertson & Saad Filho (2020).
23. There are many ways in which food is ever-present from snacks through ready-made and convenient foods through takeaways, cafes and restaurants. These are complemented by fridge-freezers that can readily be raided and rationalise bulk purchases from hype-markets, through car trips, each of which, incidentally, does not originate with financialized credit (and debt) but is fuelled by it. See Rinkinen, Shove & Smits (2017).
24. Hence the social construction of the division between the natural and the artificial, as in bottling wine, for which see Brooks and Francis (2020) in what might be interpreted as the cork SoP.
25. Indeed, ethical systems themselves can be seen as subject to the 10Cs, see Fine (2013).
26. This is a form of commodity fetishism, in which the properties of commodities are as much concealed as revealed through being provided through the market.
27. More specifically, for example, the same object of consumption, such as a (McDonald's) hamburger, can have both different material and cultural content in different situations, and these be differently determined too—as cheap food for the rich, or a treat for the poor, as emulation of, or cultural subordination to, America, as bog standard item of convenience or potentially tainted by poor working conditions and horse meat, Fine (2007).
28. This is not to suggest water implications of ever-expanding food systems are negligible as it deploys over 90% of the water used in economic activity, with meat eaters roughly indirectly responsible for twice the amount accounted for by vegetarians. By the same token, cattle are responsible for a significant contribution to greenhouse gases. Thanks to Tony Allan for these points.
29. But see also Iga (2014) who views globalization of food provisioning as disembedding consumer from the social and the natural, with localized responses leading to hybridity.
30. Although this is now complemented by authoritarian forms of freedom to choose through 'nudging' consumers to make the right choices.

31. Today, much is made of nudging as well as advising, by food product placement for example. Even if successful for some consumers, it raises the question of who buys the food if it continues to be supplied!
32. Interestingly, and with more than a tinge of irony, the financialization of food has itself been appropriately seen as, unwittingly, strengthening the omnivore's paradox of ignorance through financial displacement, as opposed to locational displacement from knowledge through active participation in the production of food, or agriculture at least. As Clapp (2014, p. 797) puts it:

    > First, a new kind of distancing has emerged within the global food system as a result of financialization that has (a) increased the number of the number and type of actors involved in global agrifood commodity chains and (b) abstracted food from its physical form into highly complex agricultural commodity derivatives. Second, this distancing has obscured the links between financial actors and food system outcomes in ways that make the political context for opposition to financialization especially challenging.

    See also reference to 'greater distancing ... by encouraging more abstraction of the commodity from its original form, in this case into a 'virtual' financial derivative product' (p. 810). Ioris (2016) refers to 'displacement, financialization and mystification' (p. 85), and Leguizamón (2016, p. 313) to 'the socio-ecological contradictions that arise from the processes of distancing and abstraction which accompany the financialization of the corporate food system under neoliberal globalization'. Note also that agro-tourism is a novel development in reuniting eating with its agricultural origins.
33. See also Guthman (2011) who points to a political economy of bulimia, one that promotes consumption while also insisting upon thinness.
34. Through participation in an ESRC research programme launched to research the (UK) Nation's Diet in order to understand why consumers did not follow healthy eating guidelines, Murcott (Ed.) (1998).
35. Alongside other syndromes such as clinically and psychologically identified eating disorders, as well as socially constructed conditions around fatness and thinness, variously interpreted, Guthman and DuPuis (2006) and Pirie (2011).
36. See Lohman and Sexton (2010).

References

Baines, J. (2017). Accumulating through food crisis?: Farmers, commodity traders and the distributional politics of financialization. *Review of International Political Economy, 24*(3), 497–537. https://doi.org/10.1080/09692290.2017.1304434.

Barrett, C., Reardon, T., Swinnen, H., & Zilberman, D. (2019). *Structural transformation and economic development: Insights from the agri-food value chain revolution*. Working Paper. Dyson Cornell SC Johnson College of Business. Retrieved from http://barrett.dyson.cornell.edu/files/papers/BRSZ%2013%20Aug%202019.pdf.

Baud, C., & Durand, C. (2012). Financialization, globalization, and the making of profits by leading retailers. *Socio-Economic Review, 10*(2), 241–266.

Bayliss, K., Fine, B., & Robertson, M. (Eds.). (2018). *Material cultures of financialisation*. London: Routledge, reproduced special issue of *New Political Economy*, 2017, 22(4).

Bayliss, K., Fine, B., Robertson, M., & Saad Filho, A. (2020). *Neoliberalism, financialisation and welfare: The political economy of social provision in the UK*. Cheltenham: Edward Elgar, in preparation.

Boffo, M., Fine, B., & Saad-Filho, A. (2018). Neoliberal capitalism: The authoritarian turn. In L. Panitch & G. Albo (Eds.), *Vol 55: Socialist Register 2019: A world turned upside down?* (pp. 247–270). London: Merlin Press.

Brisset, N. (2019). *Economics and performativity: Exploring limits, theories and cases*. London: Routledge.

Brooks, S. (2016). Inducing food insecurity: Financialisation and development in the post-2015 era. *Third World Quarterly, 37*(5), 768–780.

Brooks, A., & Francis, R. (2020). Succeeding from nature: The non-human agency of Portuguese cork. *The Geographical Journal, 186*(2), 237–246.

Callon, M. (1986). Some elements of a sociology of translation: Domestication of the scallops and the fishermen of Saint Brieuc Bay. In J. Law (Ed.), *Power, action and belief: A new sociology of knowledge? Sociological review monograph* (Vol. 32, pp. 196–233). London: Routledge and Kegan Paul.

Callon, M., & Caliskan K. (2005). *New and old directions in the anthropology of markets*. Paper prepared for "New Directions in the Anthropology of Markets", Wenner-Gren Foundation, New York, April 9.

Carolan, M. (2020). Ethical eating as experienced by consumers and producers: When good food meets good farmers. *Journal of Consumer Culture*, 1–21. https://doi.org/10.1177/1469540519899967.

Clapp, J. (2014). Financialization, distance and global food politics. *Journal of Peasant Studies, 41*(5), 797–814.

Cramer, C., Johnston, D., Muller, B., Oya, C., & Sender, J. (2017). Fairtrade and labour markets in Ethiopia and Uganda. *Journal of Development Studies, 53*(6), 841–856.

Cramer, C., & Sender, J. (2019). Oranges are not only fruit: The industrialization of freshness and the quality of growth. In R. Kanbur, A. Noman, & J. Stiglitz (Eds.), *The quality of growth in Africa* (pp. 209–233). New York: Columbia University Press.

Cramer, C., Sender, J., & Oqubay, A. (2020). *African economic development: Evidence, theory, policy.* Oxford University Press.

Dixon, M. (2014). The land grab, finance capital, and food regime restructuring: The case of Egypt. *Review of African Political Economy, 41*(140), 232–248.

do Nascimento, R., Frederico, S., & Saweljew, Y. (2019). Financial capital and land control: New rentiers on the Brazilian agricultural frontier. *Revista, 22*(50), 261–286.

Edelman, M., Oya, C., & Saturnino, M. (2013). Global land grabs: Historical processes, theoretical and methodological implications and current trajectories. *Third World Quarterly, 34*(9), 1517–1531.

Fairbairn, M. (2014). 'Like Gold with Yield': Evolving intersections between farmland and finance. *Journal of Peasant Studies, 41*(5), 777–795.

Fairbairn, M. (2015). Foreignization, financialization and land grab regulation. *Journal of Agrarian Change, 15*(4), 581–591.

Fine, B. (1979). On Marx's theory of agricultural rent. *Economy and Society*, 241–278.

Fine, B. (1993). Resolving the diet paradox. *Social Science Information, 32*(4), 669–687.

Fine, B. (1995). Towards a political economy of anorexia? *Appetite, 24*(3), 231–242.

Fine, B. (1998). *The political economy of diet, health and food policy.* London: Routledge.

Fine, B. (2002). *The world of consumption: The cultural and material revisited.* London: Routledge.

Fine, B. (2003a). Callonistics: A disentanglement. *Economy and Society, 32*(3), 478–484.

Fine, B. (2003b). Political economy and nature: From ANT and environmental economics to bioeconomics? *Journal of Interdisciplinary Economics, 14*(3), 357–372.

Fine, B. (2004). Debating production-consumption linkages in food studies. *Sociologia Ruralis, 44*(3), 332–342.

Fine, B. (2005). From actor-network theory to political economy. *Capitalism, Socialism, Nature, 16*(4), 91–108.

Fine, B. (2006). Debating the 'new' imperialism. *Historical Materialism, 14*(4), 133–156.

Fine, B. (2007). From sweetness to McDonald's: How do we manufacture (the meaning of) foods? *The Review of Social & Economic Studies, 29*(2), 247–271.

Fine, B. (2013). *Economics—Unfit for purpose: The director's cut.* SOAS Department of Economics Working Paper Series, No. 176. Retrieved from http://www.soas.ac.uk/economics/research/workingpapers/file81476.pdf. Revised and shortened to appear as, Economics: Unfit for purpose. *Review of Social Economy, LXXI*(3), 373–389.

Fine, B. (2016). From performativity to the material culture of legal expertise? *London Review of International Law, 4*(3), 477–494.

Fine, B. (2019). Marx's rent theory revisited? Landed property, nature and value. *Economy and Society, 48*(3), 450–461.

Fine, B., & Leopold, E. (1993). *The world of consumption.* London: Routledge.

Fine, B., Heasman, M., & Wright, J. (1996). *Consumption in the age of affluence: The world of food.* London: Routledge.

Fine, B., Heasman, M., & Wright, J. (1998). What we eat and why: A socioeconomic approach to standard items in food consumption. In A. Murcott (Ed.), *The nation's diet: The social science of food choice* (pp. 95–111). London: Longmans.

Fine, B., & Saad-Filho, A. (2016). Thirteen things you need to know about neoliberalism. *Critical Sociology, 43*(4–5), 685–706.

Fischler, C. (1980). Food habits, social change, and the nature/culture dilemma. *Social Science Information, 19*(6), 937–953.

Fischler, C. (1988). Food, self, and identity. *Social Science Information, 27*(2), 275–292.

Foster, C., McMeekin, A., & Mylan, J. (2012). The entanglement of consumer expectations and (eco) innovation sequences: The case of orange juice. *Technology Analysis & Strategic Management, 24*(4), 391–405. https://doi.org/10.1080/09537325.2012.663963.

Friedmann, H., & McMichael, P. (1989). Agriculture and the state system. *Sociologia Ruralis, 29*(2), 93–117.

GBD 2017 Diet Collaborators. (2019). Health effects of dietary risks in 195 countries, 1990–2017: A systematic analysis for the Global Burden of Disease Study 2017. *The Lancet, 393*, 1958–1972. https://doi.org/10.1016/S0140-6736(19)30041-8.

Ghosh, J. (2010). The unnatural coupling: Food and global finance. *Journal of Agrarian Change, 10*(1), 72–86.

Goodman, D. (1999). Agro-food studies in the 'Age of Ecology': Nature, corporeality, bio-politics. *Sociologia Ruralis, 39*(1), 17–38.

Goodman, D. (2001). Ontology matters: The relational materiality of nature and agri-food studies. *Sociologia Ruralis, 41*(2), 182–200.

Goodman, D. (2002). Rethinking food production-consumption: Integrative perspectives. *Sociologia Ruralis, 42*(4), 271–277.

Goodman, D., & DuPuis, E. (2002). Knowing food and growing food: Beyond production-consumption debate in the sociology of agriculture. *Sociologia Ruralis, 42*(1), 5–22.

Guthman, J. (2011). *Weighing in: Obesity, food justice, and the limits of capitalism*. Berkeley: University of California Press.

Guthman, J. (2015). Binging and purging: Agrofood capitalism and the body as socioecological fix. *Environment and Planning A, 47*(12), 2522–2536.

Guthman, J., & DuPuis, E. (2006). Embodying neoliberalism: Economy, culture, and the politics of fat. *Environment and Planning D, 24*(3), 427–448.

IDI. (2017). Unjust enrichment: How the IFC profits from land grabbing in Africa, outsourcing development: Lifting the veil on the World Bank Group's lending through financial intermediaries. Inclusive Development International. Retrieved from http://www.inclusivedevelopment.net/wp-content/uploads/2017/04/Outsourcing-Development-Africa.pdf.

Iga, M. (2014). Changing agri-food systems in the global economy. *Japanese Journal of Human Geography, 66*(6), 552–564.

Ioris, A. (2016). The politico ecological economy of neoliberal agribusiness: Displacement, financialisation and mystification. *Area, 48*(1), 84–91.

Isakson, S. (2014). Food and finance: The financial transformation of agro-food supply chains. *Journal of Peasant Studies, 41*(5), 749–775.

Larder, N., Sippel, S., & Lawrence, G. (2015). Finance capital, food security narratives and Australian agricultural land. *Journal of Agrarian Change, 15*(4), 592–603.

Leguizamón, A. (2016). Disappearing nature? Agribusiness, biotechnology and distance in Argentine soybean production. *Journal of Peasant Studies, 43*(2), 313–330.

Lehecka, G. (2014). Have food and financial markets integrated? *Applied Economics, 46*(18), 2087–2095.

Li, T. (2015). Transnational farmland investment: A risky business. *Journal of Agrarian Change, 15*(4), 560–568.

Lohman, L., & Sexton, S. (2010). Carbon markets: The policy reality. *Global Social Policy, 10*(1), 3–6.

MacGill, M. (2018). How soda impacts diabetes risk. *Medical News Today*. Retrieved from https://www.medicalnewstoday.com/articles/259604.

Malm, A. (2019). Against hybridism: Why we need to distinguish between nature and society, now more than ever. *Historical Materialism, 27*(2), 156–187.

Marsden, T., Lee, R., Flynn, A., & Thankappan, S. (2010). *The new regulation and governance of food: Beyond the food crisis?* London: Routledge.

McMichael, P. (2012). The land grab and corporate food regime restructuring. *Journal of Peasant Studies, 39*(3), 681–701.

Murcott, A. (Ed.). (1998). *The nation's diet: The social science of food choice*. London: Longmans.

Murcott, A. (2019). *Introducing the sociology of food and eating*. London: Bloomsbury.
Newman, S. (2009). Financialization and changes in the social relations along commodity chains: The case of coffee. *Review of Radical Political Economics, 41*(4), 539–559.
Pirie, I. (2011). The political economy of bulimia nervosa. *New Political Economy, 16*(3), 323–346. https://doi.org/10.1080/13563467.2011.519020.
Pirie, I. (2016). Disordered eating and the contradictions of neoliberal governance. *Sociology of Health & Illness, 38*(6), 839–853. https://doi.org/10.1111/1467-9566.12408.
Rinkinen, J., Shove, E., & Smits, M. (2017). Cold chains in Hanoi and Bangkok: Changing systems of provision and practice. *Journal of Consumer Culture, 19*(3), 1–19.
Rippe, J. M., & Angelopoulos, T. J. (2016). Relationship between added sugars consumption and chronic disease risk factors: Current understanding. *Nutrients, 8*(11), 697. https://doi.org/10.3390/nu8110697.
Rossman, P., & Greenfield, G. (2006). Financialization: New routes to profit, new challenges for trade unions. *Labour Education, Quarterly Review of the ILO Bureau for Workers' Activities*, No 142. Retrieved from http://www.iufdocuments.org/www/documents/Financialization-e.pdf.
Salerno, T. (2014). Capitalising on the financialisation of agriculture: Cargill's land investment techniques in the Philippines. *Third World Quarterly, 35*(9), 1709–1727.
Sippel, S., Larder, N., & Lawrence, G. (2016). Grounding the financialization of farmland: Perspectives on financial actors as new landowners in rural Australia. *Agriculture and Human Values*, online, https://doi.org/10.1007/s10460-016-9707-2.
Sommerville, M., & Magnan, A. (2015). 'Pinstripes on the prairies': Examining the financialization of farming systems in the Canadian prairie provinces. *Journal of Peasant Studies, 42*(1), 119–144.
Soper, R. (2019). How wage structure and crop size negatively impact farmworker livelihoods in monocrop organic production: Interviews with strawberry harvesters in California. *Agriculture and Human Values, 37*, 325–336.
Stevano, S. (2020). Marx and the poor's nourishment: Diets in contemporary sub-Saharan Africa. In A. Mezzadri (Ed). *Marx in the field*. London: Anthem Press, forthcoming.
Stevano, S., Johnston, D., & Codjoe, E. (2020). Better decisions for food security? Critical reflections on the economics of food choice and decision-making in development economics, *Cambridge Journal of Economics*, beaa012. https://doi.org/10.1093/cje/beaa012.

Visser, O., Clapp, J., & Isakson, S. (2015). Introduction to a symposium on global finance and the agri-food sector: Risk and regulation. *Journal of Agrarian Change, 15*(4), 541–548.

Warde, A. (2016). *The practice of eating*. Cambridge: Polity Press.

Williams, J. (2014). Feeding finance: A critical account of the shifting relationships between finance, food and farming. *Economy and Society, 43*(3), 401–431.

Wilson, B. (2019). *The way we eat now: Strategies for eating in a world of change*. London: Fourth Estate.

Wrigley, N., & Wood, S. (2018). An economic geography of globalizing retail: Emergence, characteristics, contribution. In G. Cook, J. Beaverstock, J. Johns, F. McDonald, & N. Pandit (Eds.), *The Routledge companion to the geography of international business*. London and New York: Routledge.

CHAPTER 6

Conclusion: The Contribution of the SoP Approach

6.1 Introduction

The subject of consumption has long attracted attention across the social sciences as academics have sought to understand who gets what, why, how and with what significances and meanings. These issues have recently become more pressing in light of climate change and growing inequality and impoverishment. The SoP approach evolved in the early 1990s in response to the limitations that derived in particular from siloed horizontal approaches to consumption studies from within particular disciplines. With a commitment to genuine interdisciplinarity and grounded in political economy, the SoP approach has been acutely critical of mainstream economics where consumption outcomes continue to be seen to revolve around the result of the actions of hypothetically atomistic, utility-maximizing individuals, operating in a context of more or less perfectly working markets.

The SoP approach views consumption as irreducibly attached to integrated chains of provisioning, linked to the materiality of specific goods or services and shaped by the agents within and associated with provisioning processes. Corresponding vertical chains of provisioning intersect with horizontal social structures such as gender, race and class to give rise to consumption norms which reflect socially and culturally determined propensities to consume across goods or services. Consumption (and production) is associated with specific and complex material cultures which determine and reflect the meanings and understanding that are attached, often subtly, to goods and services and to their modes of provisioning. For

K. Bayliss, B. Fine, *A Guide to the Systems of Provision Approach*,
https://doi.org/10.1007/978-3-030-54143-9_6

the SoP approach, it is all of these factors, rather than individual preferences, that drive and shape consumption outcomes.

Consumption is ultimately about who gets what but this attracts different terminologies depending on the circumstances. At one level, consumption is associated with shopping but welfare services such as health and education are also 'consumed'. Rather than distinguishing between the nature of what is consumed or who is the provider, the SoP approach takes each sector or sub-sector as a separate case. In each, the nature of the provider and the rationale for consumption will vary. 'Consumption' of education, for example, tends to be understood as a source of personal and social development benefitting both individual and society. Health services are similar, with both private and public benefits and costs.

There might be considered to be a spectrum of provisioning from public goods to fully privatized commodities. Publicly financed and provided primary education, for example, would be at one extreme (although the private sector will still be involved in one way or another in terms of services deployed). At the other extreme would be private tuition for a specific skill. But in between there are different degrees of commodification such as outsourcing of some education services or privately provided, fee-paying schools or where universities are required to be more or less self-financing. Each of these create and depend upon a specific set of social relations. Similarly, health services will be produced and consumed differently from primary health services to, say, cosmetic surgery. Systems also go beyond the public/private dichotomy to incorporate self-provisioning or informal and cooperative processes, such as in energy systems (Chappells and Shove 2000). Likewise, consumption assumes a different culture and terminology when it takes the form of activism with calls, for example, to boycott certain types of goods for political reasons. The SoP approach is sensitive to these differences in how such provisionings are realized, and the corresponding cultures to which they are attached. In being so, it is necessarily concerned with the specifics of which agents are involved and how and why and in what contexts.

This book has set out the history, the core elements and some key examples of the SoP approach as it has previously been applied. This chapter summarizes some of the main elements in SoP analyses and some of the more significant contributions emerging from the use of the approach before considering possible future directions for SoP-based research.

6.2 The SoP Approach in Context

The SoP approach was initially devised around the beginning of the 1990s in response to the perceived failings of consumption studies across the social sciences. Chapter 1 presented the evolution of the approach with its development divided loosely into three stages, starting initially with a focus on consumption norms in relation to consumer durables. The application of the approach was then widened to include food consumption, and this topic has continued to be one of the areas where SoP research has been most concentrated, see below and Chap. 5. A third phase of SoP studies has evolved to explore consumption and production in relation to areas where these have been primarily in the public domain. These studies took the SoP approach into the realm of social policy.

The SoP approach draws on different academic disciplines (Chap. 2). There are parallels with Marxist political economy (MPE) in the attention to the social relations that underpin the production of apparently free-standing commodities. Chapter 2 shows how the strands of different theories can be brought together to forge a framework for the fundamentals of SoP analyses. The core elements are agents, structures, processes, relations and material cultures. Material cultures are located in the relationships between people and things (see Chap. 3). Such cultures are often so deeply ingrained that they can be difficult to identify. But material cultures vary depending on the context, and they are malleable. Those who shape material cultures wield considerable power as reflected, for example, in the notion of Foucauldian governmentality. But humans are reflective, and cultures will be interpreted and impact differentially. To help to navigate this complex terrain, Chap. 3 provides a grouping of the factors that shape cultural systems under ten headings (known as the 10Cs).

As with SoPs themselves, context is also important for understanding the SoP approach. It did not emerge from thin air but builds on and contributes to a wealth of scholarship seeking to unpack what it is that leads to haves and have nots and the spectrum in between. The SoP approach is not alone in rejecting mainstream economics' preoccupation with rational utility maximizing individuals nor, at the opposite extreme, free floating attention to the discursive symbols, signs and discourses of postmodernism. There is a widespread understanding, with longstanding traditions, that consumption is a social phenomenon, and increased efforts have been made to try to understand its drivers. Decades of consumption research has shown that individual preferences and choice are only a small part of what motivates patterns and norms of consumption (Hansen 2018a), and such preferences would themselves have to be explained rather than taken for granted.

The SoP approach had become a major influence in consumption studies by the end of the 1990s, alongside a number of other approaches such as the 'circulation of culture' and 'consumption as practice'. Practice theory is closely related and complementary, to the SoP approach, taking 'practices' rather than individuals, citizens, societies, social groups or sociotechnical systems as the unit or focus of attention (Shove and Walker 2010). Practice theory is grounded in understanding the everyday and mundane aspects of consumption such as DIY.[1] Shove and Walker (2010), for example, use this approach to understand shifting patterns in bathing practices and congestion charging. In common with the SoP approach, they consider how consumption patterns change and they point to the role of agents, for example the manufacturers of products such as shower gels, in shaping these. In food consumption, Alan Warde has made a significant contribution to the practice theory literature, especially around eating out and family meals (e.g., Warde 2016).

SoP, then, was one of a number of approaches, which were seen as forging a new way of understanding consumption in which material and cultural relations are taken to be symbiotic and grounded in activities beyond the acts of consumers and consumption themselves. In contrast to subjective and individualistic approaches (somewhere between the infinitely inventive postmodernist and the predetermined individuals and goods of neoclassical economics accounts), the SoP approach was part and parcel of establishing what can be termed a material culture of consumption.

Significantly, these various approaches to the material culture of consumption were generally mutually 'friendly' on personal and intellectual levels, drawing upon and reinforcing one another and, to a large degree, establishing a strong hold over the field.[2] Thus, for example, the SoP approach features in the online handbook titled *Consumption, Everyday Life and Sustainability* (https://www.lancaster.ac.uk/fass/projects/esf/introreader99.htm) as part of an interdisciplinary approach to issues of consumption, everyday life and sustainability.[3] Chapter 5 of the handbook:[4]

> reflect[s] on the benefits of adapting a "systems of provision" framework … as a means of understanding how demand for energy, water and a range of associated utility services is created and managed … that commodities and services find their way to consumers in different ways depending on how

specific systems of provision are configured. It is the system of provision that unites a particular pattern of production with a particular pattern of consumption.

Then, and subsequently, researchers have often drawn on more than one approach. As Jackson, Russell & Ward (2004, p. 4) put it in the context of food, with the SoP approach prominent, and in welcoming the constructive chaos of different perspectives for allowing for different insights:[5]

> The recent rate of scholarly output can make it difficult to give shape to the emerging debates and conceptual fault-lines. Nevertheless, we might tentatively identify the following three trends. *First* is the 'consumption turn', where … the "filière-commodity systems-agroindustrial approach of the 1980s" has been drawn upon and developed to accommodate a greater interest in consumption, most notably through concepts such as 'systems of provision' … A *second*, and related, trend is a 'cultural turn' associated with increasing interest in the production and consumption of meanings and narratives around food, nicely captured by … [the] point that, especially in the global North, "most food is sold with a story" …. *Third* has been a 'quality turn', particularly associated with the development of alternative agro-food networks, the rise of quality assurance schemes, and strategies to valorise local and regional food products.

From the time that the SoP, and other related, approaches attained this position of prominence, they have continued to prosper as will be seen below in case of the SoP approach in particular, but to have been complemented by other approaches as well, without being eclipsed as such.[6] There is a sense in which those initially pushing so strongly in different ways for a material culture of consumption were a victim of their own success in that their approaches became so widely accepted that in some cases they were able to be adopted as second nature without the need to acknowledge, or possibly even be aware, from where adopted analytical framings originated.[7] This is most obvious, although influence is hard to prove, in the studies that understand consumption systemically, which often come close to deploying the terminology of the SoP approach if not explicitly referencing it, whilst also recognizing that systemic analysis is, unevenly across topics and fields, widespread in both scholarship and popular discourses, in the spirit of the SoP approach or otherwise. Indeed, there were good reasons why studies of consumption, and other topics too, should have leaned towards systemic understandings over the past few decades. In general, this is because of the increasingly unavoidable presence of systemic factors, however understood, such as globalization, neoliberalization and financialization as well as an equal concern with their effects such as growing inequality and impoverishment. More specifically in case of consumption, such considerations were equally linked to shifting levels, forms and cultures of provisioning (consumption norms for the SoP approach) in light of the shift from public to private provisioning through privatization and greater reliance upon a market ethos and individualized responsibilities. A leading example of the imperative of the systemic has been growing attention to the environment, for which the SoP approach has been adopted, see below.

The SoP approach, then is part of a larger movement towards systems-based understandings of consumption and of social outcomes. Apart from forging a common approach from different perspectives and with different subject matter and case studies, even if with food prominent and the everyday as opposed to the exotic to the fore, there were other reasons for this cosy relation between the various approaches. One is they were often derived from different disciplinary origins but with a strong commitment to interdisciplinarity, not just across social sciences but also one which includes applied physical sciences, not least because of interest in transport, energy, waste, the built environment (see below on sustainability). And another is the commitment to place consumption in its historical and social contexts.

This coalescing of like-minded, but differentiated, consumption approaches paves the way for a certain eclecticism, in which the SoP approach finds a place, as is evidenced by Wheeler's (2018) study of the moral economy of ready-made food (convenient but bad to eat?) drawing upon the SoP approach but also a range of other perspectives from within sociology in particular (performativity, rioting, etc).[8] She concludes (p. 1289):

> Any attempt to understand the moral economy [of ready-made meals] must pay attention to all three layers of this analytical framework (state regulation of the market, collective customs and lay normativities) and how they are instituted within a distinct instituted system of provision.

Add in provisioning itself, and this is an invitation to the 10Cs! As an early contribution in this vein of mixes of approaches, Mansvelt (2005) warmly embraces the SoP approach as it 'provided valuable insights into commodity trajectories in spatial and social contexts … in which commodities acquire form and meaning' (p. 110), thereby demonstrating 'how relations of connection differ depending on the commodities and the structures and histories in which they' are embedded (p. 124). But equally, Mansvelt rounds up other approaches to consumption—global chains, circuits of culture, Actor-Network Theory, and so on, to add their own insights.

Some have pointed to the need to supplement, or complement, the household practice approach with SoP analyses, such as Foden, Browne, Evans, Sharp & Watson (2019) in relation to social practices in the disposal of fats, oils and grease. Bell, Judson, Bulkeley, Powells, Capova & Lynch (2015) conduct an investigation of the household practices that lead to electricity consumption. They raise issues that would be found in SoP analyses, for example, that, the gendering of household consumption involves Conventions, Capacities, Rhythms, Economies and Structures.[9] They consider that, in addition to the measuring of household energy use, 'such metrics need supplementing by knowledge about practices in which electricity is embedded and their relation to systems of provision and connection to the enactment of sociality in households' (p. 98). Chappells and Shove (2000) in their 'consumption as practice' similarly draw on the SoP approach and call for it to be extended to take account of the vast array of devices and practices that mediate energy consumption. They show how energy consumers are locked into relationships of 'organisational dependency' (p. 40) such that technologies and devices can dictate householder and provider activities. They propose a framework of socio-technical system of 'scripts' and show how conflicts can arise between these.[10]

More generally, Hansen (2018b) mindfully draws upon the SoP approach (original emphasis):

> to incorporate the whole process *before* acquisition in order to understand consumption, in other words a consumer object's backward linkages.

Hansen however raises a long-standing critique of the SoP, and other systems-based approaches, that, while they purport to be concerned with consumption, the real focus is on production, rather than the linkages. As Goodman and Dupuis (2002, p. 7) state: 'The consumer emerges only to disappear again into a production-centred framework.' Hansen sees this as raising a 'theoretical conundrum' in how to connect production processes with a holistic account of consumption and everyday practices. He suggests that practice theories need to be analyzed 'through macro-scale systems of provision and political economic frames and conditions for practices'. He stresses the importance of material cultures:

> without an empirical grounding in the actual 'doings' of the people consuming, in the everyday practices consumption takes place in, macro-level approaches provide a shallow reading, stripped of the ability to understand the multifaceted meanings and drivers of consumption.

Thus, the SoP approach dovetails with other systems-based approaches, to strengthen understandings of the drivers of consumption, the material practices that underpin these and the ways in which these are situated in broader contexts.

The emergence of a portfolio of systems-based approaches has lent itself to addressing a related but distinct portfolio of themes, some of which are mentioned in the previous chapters. SoP research has featured strongly in studies of food consumption (Chap. 5) and increasingly in sustainability (see below). But the approach has been widely used in different and diverse contexts. Examples include fast fashion and denim jeans (Brooks 2015a, b); bread in South Africa drawing on the 10Cs in the context of the industrialization of production at the expense of consumer concerns for price and quality (van der Walt 2016); consumption and supply chains for orange juice (Foster, McMeekin & Mylan 2012). Williams, Spotswood, Parkhurst & Chatterton (2019) examine (local) transport and sustainability from a SoP perspective; Hansen (2018a) deploys the SoP approach to understand the 'meatification' practices of consumer choices in Vietnam, especially with the shift to large-scale retailing. Rinkinen, Shove & Smits (2019) look at the role of fridge/freezers in shifting patterns of consumption. They link food consumption with environmental sustainability. They find that eating patterns in Bangkok and Hanoi have been transformed by the pivotal role played by the widening use of fridge freezers and the 'cold' supply chain. This in turn has had an effect on energy demand. They call for (p. 394) 'more situated understandings of consumption and of the extent to which increasingly resource intensive ways of life are inextricably and perhaps unavoidably embedded in urban development'. Serrano and Brooks (2019) show that a surge in demand for avocados in the global North has created new systems for consumption and production. Their study of farming in Colombia indicates that production and exports have been booming. However, small farmers that lack connections to international supply networks and access to large scale farming methods have been squeezed out.

Chapter 4 explores in more detail some examples of how the SoP approach has been applied in practice, drawing on studies related to the establishing of consumption norms (in housing) and the (often less visible) SoPs which underpin everyday life (in water, health services and clothing). The studies cited reveal that mundane activities such as turning on the tap appear to be unchanging over decades but such superficial continuity obscures major changes in the underlying social relations. For example, some water providers are now owned by unknown offshore financial investors. In health services in England, these continue to be mostly free at the point of delivery but this conceals a wealth of sector reform which in some cases created market opportunities for private business. More significantly the health service sector has been crushed by lack of funding after a decade of government austerity policies which has created a narrative that the public health service is unaffordable and inefficient. These case studies demonstrate the importance of material cultures in mediating shifts in the SoP and of the power relations that work through into consumption narratives (see Chap. 4 for more details).

These (and other) SoP studies highlight that consumption outcomes result from the intersection of numerous elements in the SoP. Consumer choices do not emerge spontaneously out of some kind of innate set of preferences but are changeable and shaped by those with an interest in outcomes. But the cases highlight the importance of specificity. An ever-present challenge in conducting SoP research is that it is not clear from the outset where one SoP ends and another begins. Ultimately everything connects to everything else. Boundaries cannot be set in advance but emerge inductively as part of the research process. These complexities are not easily resolved and require inductively framing SoPs in terms of interactions between them according to the nature of the SoPs themselves and how they interact with one another. And, of course, there is the issue of intersections within what may or may not be treated as separate SoPs, for example, when it comes to public and private housing (Moore and Dalton 2016).

The intersections across SoPs and the connections between the agents in the provisioning systems means that the SoP approach has increasingly been used to finesse the relationship between the macro and the micro, the social and the specific. The macro/micro issue is indicative, as are many of the other contributions already covered, of a growing weight of studies that engage with intersectoral aspects of consumption even where a particular sector may be of initial focus, as with the concern of Iacovidou, Millward-Hopkins, Busch, Purnell, Velis, Hahladakis, Zwirner & Brown (2017) with the recovery of resources from waste in the circular economy. More generally, Bell, Judson, Bulkeley, Powells, Capova & Lynch (2015) appropriately refer to a system of systems (of provision). For the Energy Policy Group at Exeter University, the systemic is linked to locked-in structures, relations, agencies and practices of provision[11]:

> Moreover, change occurring within each system has the potential to negatively impact upon other systems of provision.

Intersections across SoPs are also evident when it comes to health. The factors that shape health outcomes are strongly linked not just to health services but also to the SoPs for food and for housing. Thus, to address the social determinants of health, attention should be paid to the factors that create unhealthy food consumption and poor housing. Case and Deaton (2020) demonstrate this in relation to health indicators in the USA where rent seeking and discrimination in the American health system have recently led to reductions in life expectancy for those without a college degree.

The macro/micro relationship has also featured in infrastructure where the SoP approach connects macro policies regarding the financing and provisioning systems with the micro activities of producers and consumers. This is demonstrated in a major research programme (involving engineers and social scientists) on UK infrastructure, *IBuild*.[12] The project deployed the SoP approach to understand better how (and what) physical infrastructure might be provided, how its separate components might be integrated, and how its potential can be best harnessed to use—infrastructures are, after all, systems of, and involved in, provisioning rather than simply being physical things,[13] and see Brown and Robertson (2014). Macro and micro also intersect in a component of the ESRC-funded *Rebuilding Macroeconomics* titled *Trajectories of Infrastructure Financing and Macroeconomic Policies in Practice*. This interdisciplinary project applies the SoP approach to infrastructure financing in the UK to understand the cultures and discourses across diverse sets of agents in shaping financing policies and outcomes.[14]

The links between micro and macro extend further, not least in contributing to the understanding of the grand concepts such as neoliberalism and social reproduction. This is so of Bayliss, Fine, Robertson &

Saad-Filho (2020) on neoliberalism, Fine (2020a) on social reproduction and Bayliss, Fine & Robertson (Eds.) (2018a) on the material culture of financialization—itself spanning the meaning of financial activity to its participants, how they assess their well-being in the wake of the GFC, what is and is meant by financial inclusion and financial literacy, and media representations of finance, quite apart from more standard SoP accounts of the impacts of financialization on (the cultures) of water and housing provisioning. This is also indicative of the extent to which the different elements of the SoP approach—those relating to norms and culture through the 10Cs, for example—can be deployed more or less on their own account, as in Fine's (2013a) discussion of the (hidden) ethics of mainstream economics.

SoP studies have highlighted that the roles of the state may be different and even contradictory across different state institutions. But, especially in the context of privatization, it is the spread of the SoP approach with increasing attention to the continuing role of the state, in social and economic infrastructure and beyond, that has attracted the use of the SoP approach, for locating PPPs in the current phase of neoliberalism for example, Fine (2020b). Significant in the SoPs where privatization has occurred has been the extent of the state's role in circumstances where it was intended to be rolled back. Private markets have required substantial, even increased, state intervention.[15] The SoP approach has through detailed case studies demonstrated the complexities and conflicts engendered by the continuing and multifarious roles of the state, and this has been one of a number of increasingly important issues for the SoP approach, not least through attention to social reproduction, (environmental) sustainability, and consumer politics.

6.3 Future Directions

As stated the SoP approach has been applied to diverse sectors and subjects. Chapter 1 sets out three significant phases in the evolution of the approach so far as reflected in key case studies of consumptions norms around consumer durables, followed by food and, then, public goods and services. This section reviews three more areas where SoP analyses have made, or have the potential to make, a significant contribution.

6.3.1 *Social Policy and Social Reproduction*

SoP studies have shown how the systemic features of provisioning contribute to and perpetuate inequalities across societies. Mainstream approaches to social policy have focused on categories of welfare policies, as in the Welfare Regimes Approach (see Chap. 4). The SoP approach in contrast focuses on the specific empirically grounded study by sector and location. The approach looks beyond welfare and takes into account inequalities that are embedded within provisioning systems themselves.

The evidence gathered from SoP research has been instructive in highlighting pervasive dysfunctional effects of neoliberalism and financialization. In the UK housing sector for example, immense housing wealth sits alongside rapidly rising homelessness. This can be attributed to some degree to financialization. But the link is not straightforward. The SoP approach requires attention to why the outcomes took the shape that they did in that specific context. In the UK, financial deregulation combined with a poor supply response (for a number of reasons—see Chap. 4) and a shift in material cultures which heavily promoted homeownership. The SoP approach explores all of these and their intersections in leading to rapidly increasing house prices, creating wealth for some and putting homeownership out of reach for others.

Homeownership is part of a broader neoliberal project of individualized asset-based accumulation to replace collectivized welfare provision, with a greater reliance on the market and encouraging individuals to bear more risk. Growth in owner occupation is crucial to this. Yet the result has been a sharp increase in intergenerational inequality and its causes are evidenced in SoP analyses. In housing in the UK for example, as house prices have escalated in some parts of the country, there is a collective experience of exclusion. Meanwhile, in common with developments elsewhere in Europe, but to an extreme degree in the UK, the flipside of growing owner-occupation is growth in unaffordable private renting for those excluded from, and rapidly increasing housing benefit as a result matched by a decline in support for social housing provision. Such are the new housing consumption norms in levels and forms of provisioning!

The SoP studies demonstrate the role played by neoliberal cultures in some cases, particularly with the emphasis on individualized rather than collective provisioning. This has had differential effects across the SoPs studied. In pensions, for example, financialization relies on the conceptualizing of a pension as an individual activity rather than collective provision for the aged other than as (large and disadvantaged) residual. There is then a tendency to target the wealthy. The other side of individualism is residualism where the state is left to cater for those whom, for whatever reason, the individualist structures failed to be successful, such as those without housing, pensions, health insurance and so on. In Europe, there has been a shift from defined benefit schemes to defined contribution schemes so that the scheme holders bear all the risk with their pension income entirely dependent on returns in the financial markets. Meanwhile, increased financialization in the way in which pensions are managed can lead to greater systemic risk (e.g., see Bonizzi and Churchill 2017; Churchill 2013, 2018, 2019; Saritas 2014, 2016, 2020).

While the SoP approach is highly context-specific with attention to a specific sector, comparative studies demonstrate the cross-sector impact of horizontal structures and processes. Bayliss and Mattioli (2018) explore the SoPs for water, energy and local bus transport in Britain. These have evolved since privatization in the 1980s and 1990s to create structures which generate secure returns to shareholders while services are unaffordable for many and non-existent in the case of bus services that were cut because considered commercially unviable once privatized. Individualism is evident in the changes in the way that water and energy is consumed, with an increased emphasis on metering. This means that the individual is responsible for monitoring consumption and thereby ensuring affordability. When it comes to transport, some individuals who are no longer served by bus services after privatization are forced into car dependence. While the developments are variegated across sectors, cross-cutting drivers such as neoliberalism and financialisation are shared across these essential services, in some cases associated with multiple deprivations.

Neoliberalism, with the framing of the world in terms of market interactions, is successful in distancing consumers from the underlying system of social relations that underpin production and consumption. With such narratives, attention is focused on the commodity and the consuming individual. The SoP approach, however, provides a framework to connect end users with the relations and agents in the provisioning chain that are mostly hidden from view, such as the extractive practices of financial investors in the water sector, or the opaque trade in cast off clothes from wealthy countries which are sold to developing countries, or the unequal allocation of value from the sale of clothing across retailers and agricultural workers (see Chap. 4). Such analyses are valuable in highlighting the structures and processes by which inequality evolves and is sustained.

6.3.2 *Sustainability and SoP*

System-based analyses have long been a feature of environmental debates (e.g., see Meadows (2008) which draws on decades of environmental systems-based work by Donella Meadows). As the climate crisis has become increasingly urgent, the SoP approach has proven to be well-placed to facilitate understandings of consumption patterns. Particularly prominent over the recent period have been studies of the environmental system(s) and sustainability, across diverse applications and contexts, applying practice theory and SoP and sometimes both. For example, the SoP approach plays an important role in an on-going research project at the University of Leeds, titled 'Living Well Within Limits'. The project links biophysical inputs with social outcomes and considers how the modes of provisioning create structures that lead to specific outcomes in terms of resource use, for example, with transport and energy systems.[16]

The SoP approach has featured in a number of research papers related to environmental sustainability. For example, Bulkeley and Askins (2009) examine waste and everyday practice.[17] The issue of resource recovery from waste systems is addressed by Iacovidou, Millward-Hopkins, Busch, Purnell, Velis, Hahladakis, Zwirner & Brown (2017), combining science and engineering with a socio-political analysis to develop a system-wide perspective. Consumption and sustainability in general is considered by Reisch, Cohen, Thøgersen & Tukker (2016), seeking to match systemic sustainable provisioning alongside other goals. At the other extreme of detail, use of firewood in Finland is examined by Rinkinen (2013) in response to energy shocks, and there is consideration of energy systems for new housing construction in the UK by Monahan and Powell (2011). For Judson, Bell, Bulkeley, Powells & Lyon (2015), for heat pumps and energy efficiency in social housing, no explicit reference is made to the SoP approach but one is sought that offers 'a perspective that unites all elements of energy production, distribution and consumption under the single concept of a *system of provision*' (p. 27).[18] Sustainable provisioning of food is addressed by Kantamaturapoj, Oosterveer & Spaargaren (2012) for the Bangkok middle class through supermarkets, and for carrots around the contested notion of the 'organic' by Seyfang (2007). Watts, Ilbery & Maye (2005) examine 'alternative systems of food provision', but again without reference to the SoP approach as such! And Werna (1998) looks at different systems of public sector service delivery across Nairobi, whilst the European Commission's Cordis project examines EU country consumption and sustainability across sectors such as ready-made food, broadband installation, and waste.[19] De Feijter, Van Vliet & Spaargaren (2019) use the

SoP approach to unpack the structures, processes, relations and agents in the SoP for the physical upgrading of existing old urban neighbourhoods to reduce carbon emissions (i.e. retrofitting) in China and the Netherlands.

Practice theory has been applied to link transitions in energy consumption with the increased dependence of modern societies on technological systems. Thus, infrastructural transitions are indelibly linked to devices and appliances that are embedded in practices as 'things in the background are of necessity tied to things in the foreground and to the ongoing mobilization of things in action' (Shove 2015, p. 167). Shove (2015) points out that the production and circulation of goods and commodities is embedded in social practices. Therefore, the global transportation of goods 'should not be interpreted as expressions of macro-economic and political forces or the circulation of capital or outcomes of multilevel or any other kinds of transition as if these were somehow detached from the realm of social practice' (Shove 2015, p. 167). Practices intersect with SoPs to create specific outcomes. Indeed, this, as is much more of the previous discussion, is illustrated by the wide-ranging study of how to shift transport systems to be more sustainable. As Mattioli, Roberts, Steinberger & Brown (2020, p. 1) suggest:

> Five key constituent elements of what we call the 'car-dependent transport system' are identified: i) the automotive industry; ii) the provision of car infrastructure; iii) the political economy of urban sprawl; iv) the provision of public transport; v) cultures of car consumption. Using the 'systems of provision' approach within political economy, we locate the part played by each element within the key dynamic processes of the system as a whole. Such processes encompass industrial structure, political-economic relations, the built environment, and cultural feedback loops. We argue that linkages between these processes are crucial to maintaining car dependence and thus create carbon lock-in.

Further, the private automobile is heavily supported through the provision of road infrastructure with corresponding cultural consequences for private motorists, 'These hidden subsidies, however, are often so prevalent that they can seem normal and fair', p. 16, whilst they can also be rationalized for bringing (regional) development. Mattioli, Roberts, Steinberger & Brown (2020) link production processes and elements of the SoP with practices and cultures. Scale economies feature highly in car production so there is an inherent tendency to overproduce. This then feeds into consumption with product differentiation promoted by advertising which is also heavily gendered. Furthermore, driving has a high cultural status which means that motorists have political importance although narratives around driving are deeply ingrained, with pro-car decisions linked to economic growth such that car dependence is seen as apolitical.

In short, both infrastructural and cultural elements are subject to mutually reinforcing 'lock-ins'. But they are equally subject to resistances and the struggle for alternatives in terms of public transportation systems. This raises the issue of consumer politics.

6.3.3 *Consumption, Consumer Politics and Activism*

The SoP approach is inevitably drawn into consumer politics, shedding light on both its nature and hence its limitations. As such, consumer politics is a broad term covering concern over monopoly pricing, quality and price of goods and so on at one extreme, to the formation of consumer cooperatives and campaigning over conditions of provisioning at the other. As Fine (2005, 2013b) argues, consumer politics has self-limitations in light of its nature insofar as the issues themselves are limited to the immediate confines of the consumer and the consumed. After all, we all consume and all are (individual) consumers. But, insofar as the determinants of consumption are traced back to their origins in production, the issues involved become much deeper sources of conflict, over wages and conditions, child labour, uneven development, and environmental degradation. As consumers we are all engaging with the social relations of production, however distantly, and these are liable to be exploitative. The purchase of an avocado at a supermarket in Europe is to buy into global chains of production but relations in global food systems are simply unseen (Serrano and Brooks 2019). As Brooks (2015a, p. 8) writes in relation to fast fashion, the act of consumption 'makes us complicit in a system which is denying people in the global South the chance to escape poverty'.

As knowledge of production systems become more widespread, consumer politics may trigger campaigning but the stronger and deeper it becomes, the more likely it is to be transformed into contestation over something else with more or less clear 'battle-lines' of conflict, with producers, the state or otherwise. The concerned consumer (for own price and quality of what is consumed) becomes the concerned citizen and consumer politics is stripped of, or supplemented by, more general concerns that are of necessity engaged in conflict with those who are provisioning.

Further, to the extent that the individual consumer engages with others more or less campaigning organizationally, as in consumer boycotts for example (so becoming less consumer and more politics) and especially for collective, non-market consumption (as with social policy and provisioning), then consumption tends to be transformed, beyond terminology alone, into struggle for what is no longer deemed to be consumption as such but for both political change to a greater or lesser extent and/or for welfare provision in detail and overall as the welfare state. Significantly, with the turn to privatization in all its forms, including the new public sector management, there is the corresponding shift in the material culture from welfarism/citizenship to individuals as clients/consumers, redefining how welfare is considered as well as what is provided and how.

Brooks (2015a) considers the role of consumer-oriented campaigns to improve the extreme inequalities embedded in global production and consumption of clothing. He shows that ethical consumption, such as Fairtrade, brings morality to the market. Fairtrade labour may attract a slightly higher wage or improved conditions but the underlying capitalist social relations are unchanged. Rather than alleviating inequality, he sees so-called ethical consumption as capitalism's 'elegant solution' (p. 218) to the criticisms raised, and one which serves to embed the 'logic of inequality' (p. 218). In the same vein, clothing retailers are promoting recycling to provide an apparently virtuous processing of unwanted clothing that also makes space for more consumption. Individual adjustments to consumption of clothing are not leading to widespread change in the global sector. And nor is any dramatic realignment expected from shifts in consumption patterns in developed countries as 'there is no real political appetite for transformation that could disrupt everyday life' (Brooks 2015a, p. 252).

Thus, the SoP approach highlights the extensive and complex channels which underpin specific social, environmental and economic outcomes, and can therefore be helpful in identifying 'leverage points' for interventions and pressure for change (Pirgmaier and Steinberger 2019). But it is also instructive in pinpointing the challenges involved. Transformational change is likely hindered by a reluctance to relinquish home comforts and is also impeded by deeply embedded narratives and structures that lock in the status quo. In addition, the SoPs themselves as well as their core elements are often under the radar. Core systems of extraction and the cultures by which these are sustained have become invisible. For example, the essential systems on which many societies depend (such as housing, water, energy, transport, food and health) are largely absent from mainstream political and economic discussions (Schafran, Noah Smith & Hall 2020). Schafran et al. (2020) apply the SoP approach to derive the concept of 'spatial contracts' which builds on the notion of social contracts. They call for greater attention to these essential, and highly context-specific systems, which enable societies to function.

6.4 Conclusion

A century ago, initially prompted by concerns over the quality of the health of recruits to fight in the Boer War, and reinforced by the same conundrum going into the First World War, Winter (1980, p. 211) observes in retrospect that:

> **No observer of political debate in** early twentieth-century Britain could have failed to notice the frequency with which the theme of the relationship between Imperial power and public health was discussed. In the aftermath of the Boer War, there was an avalanche of speculation and gloomy prognostication about the causes and likely outcome for the British nation and Empire of the supposed physical deterioration of the British male population. *What exercised many politicians and military men was the fact that between forty and sixty percent of recruits for the British Army were turned down as physically unfit for service.* (second emphasis added)

By coincidence or otherwise, the globe was soon to be hit by the Spanish flu pandemic and, possibly, the death toll was worsened by those weakened by chronic under-nutrition and acute wartime conditions.[20]

Jump forward fifty years and we found ourselves, or the USA at least, in the midst of the Vietnam War. At that time, no one was aware of the (US) epidemic of reduced life-expectancy with which it was going to be associated due to suicides and drug addiction, (Case and Deaton 2020) but there was a keen awareness of the *under*-nutrition of many US citizens. As reported in its fiftieth anniversary retrospect, Mande, Willett, Auerbach, Bleich, Broad Leib, Economos, Griffin, Grumbly, Hu, Koh, Mozaffarian, Pérez-Escamilla, Seligman, Story, Wilde & Woteki (2020, p. 5) recall:

> The 1969 White House Conference on Food, Nutrition, and Health was a landmark event. Commissioned by President Nixon and chaired by Dr. Jean Mayer, the historic 1969 conference convened a diverse group of organizations and dedicated citizens to craft a bipartisan agenda for ending hunger and malnutrition in the U.S. The meeting itself, and the report that emerged

> as a result, had significant and lasting policy impacts: subsequent years saw the expansion of the Food Stamp Program and the National School Lunch Program, the creation of the School Breakfast Program and the Special Supplemental Nutrition Program for Women, Infants and Children, and the development of dietary guidelines, nutrition education, and standardized food labeling, including the Nutrition Facts label. Of the approximately 1800 specific recommendations generated by the conference, an estimated 1650 were later implemented.

But the 2019 fiftieth anniversary of the White House Conference was not marked by celebrations. The reason for this is as follows:[21]

> Today, a half century later, the U.S. faces a very different set of nutrition challenges: epidemics of diet-related obesity, diabetes, and other chronic diseases; widening disparities in food access and affordability; food insecurity; and tremendous stresses to the environment, including threats to soils, waterways, oceans, and climate. The burden of chronic illness reduces quality-of-life and life expectancy for millions of Americans, and results in healthcare costs that are placing unprecedented strain on the budgets of federal, state and local governments, businesses, and families. Overweight and obesity are threatening our national security by disqualifying young men and women from military service. Meanwhile, environmental degradation caused by global food production is exacerbating climate change and depleting natural resources.

And nor is potential for war nor pandemic off the agenda. On the one hand, it is reported that, obesity is affecting national security in the USA with 71% of young people between the ages of 17 and 24 failing to qualify for military service and obesity disqualifies 31% of these.[22] These ineligibility rates are a major reason why the US Army has not been on track to meet annual recruitment goals (Maxey, Bishop-Josef & Goodman 2018).[23]

On the other hand, although it could hardly have anticipated what was occurring when publishing its Report in March, 2020, it was accompanied by the Covid 19 pandemic, one at least as serious as its counterpart a century previously. And, whilst knowledge of the current virus remains limited and evolving, its incidence and severity on those with diabetes is known to be particularly harsh[24]—accounting for one-third of all recorded hospitals deaths as such in the UK at time of writing.

In some sense, the spread of, and response to, the virus unfortunately offers rich analytical pickings for the SoP approach. The diversity of impacts across countries is testimony to the importance of context and the differences in structures, processes, relations and agents as well as material cultures in each country, as opposed to imposing the added value of one-or-more-size-fits-all models. Compare, for example, the structure and preparedness of health systems. Yet this alone is not an indicator of outcomes with some countries, such as Greece, achieving remarkably low rates of mortality despite a health sector that has been heavily weakened by decades of austerity. Outcomes are shaped by the intersections of health services with numerous factors from demographics to political regimes and structures of class, race and gender. As emphasised, diet, itself shaped by the components of the SoP for food (Chap. 5), is a cause of some of the major underlying health conditions that increase susceptibility to the virus such as obesity, heart disease and diabetes.[25] No doubt housing, poverty, working conditions and other indices of disadvantage will come to the fore alongside the availability of data in pinpointing the trajectory of the virus.[26]

As this book was nearing completion, then, in June 2020, the world was in the grip of the coronavirus pandemic. The response and impacts within and across different countries have been unprecedented and diverse. They have had their counterpart in academic and more popular outpourings in terms of causes and consequences, cures and lessons. Within progressive commentary in particular, this has proven an ideal opportunity for engaging in generous helpings of I told you so—whatever had been the previous takes on an era of globalized, neoliberalized, financialized, inegalitarian, intersectional austerities, the pandemic could be read off as its inevitable result with more oppression, impoverishment and exploitation in its wake.

We do not doubt there is much to be learnt from such analyses although we also suspect many of these will not outlive the pandemic itself.[27] It would also be natural to close this book by opportunistically jumping on the corona bandwagon ourselves—even to posit, by way of parody, one or more virus SoPs around who gets it and what gets done about it. We do, though, have no such inclinations although we do insist on the role that the SoP approach can play in understanding the virus, the responses to it, and even ways forward afterwards. Rather, our understanding would be, first, both to ground the pandemic within the context that has preceded it—the global, neoliberal, and so on—as well as to insist upon the politically determined, and often unpredictable, ways in which corresponding relations, structures, agencies and processes have not so much been followed through as suspended.

Second, this suspended diversity is a consequence of an enduring feature of the contemporary neoliberal state, both its high degree of centralized, interventionist governance (unevenly deployed) and its strong susceptibility to, even reliance upon, political populism (and more or less concessions to controlled devolution). The government's slogan in the UK was that it would do whatever it takes to beat the virus (even if late with lockdown, protective clothing and testing) and, indeed, its measures have been extraordinary, and often commendable, in terms of levels of expenditure and support to working people. Almost inevitably, those constituencies that have suffered the most from interventionist neglect in the pandemic are those that did not readily fall within the scope of the centralised institutions of governance and its reach to lower levels. Most notable have been the residents of (and workers in) care homes that had been a site of residualized, commercialized austerity over many years, quite apart from frontline workers who must have been astonished to find themselves in the forefront of popular approval and sympathy (even if without adequate protection).

Third, as indicated, if to reiterate, highly unevenly within and across countries, the suspension of 'normality' for whatever needs to be done, to varying degrees across countries, has offered salutary lessons in how much the state continues to retain power and has the capacity to use it for directed goals. This includes monitoring and testing for the virus to contain spread, alongside measures on social distancing, massive expansion and diversion of health resources for the treatment of those who become seriously ill, and economic and social measures in support of those who have been deprived of their livelihoods irrespective of whether victims to the virus themselves or not. All of this has been imperative in the biggest suspension of them all, that of neoliberal ideologues from which, at least initially, there has scarcely been a whimper of protest against the massive resurrection of the interventionist state—although death by lockdown economy as worse than by virus has been widely aired and inevitably gained traction as the pandemic unevenly slackened.

Fourth, to get whatever it takes done has unwittingly made the case for at least the methodology of the SoP approach. For, whether it be medical supplies, treatments in hospital, testing, or searches for vaccines or drugs, efforts to ramp up provisioning have been forced to look at and restructure the vertical systems and the ethos to which they are attached (with the profit motive being both observed and breached to suit). On the much broader canvas of intersectional social reproduction, as the statistics emerge of the incidence and impact of the disease across age, region, other socio-economic and socio-cultural characteristics, it will become increasingly clear how not only poverty, gender, race, unemployment and so on are vital sources of inequalities but that they impinge through food, housing, education, transport and health systems and so on.

Fifth, this is our second major crisis in little over a decade. The GFC after only a momentary breeze of reflation to sustain economic activity gave rise to a continuing period of austerity in the UK and elsewhere with correspondingly limited economic and social progress in deference to the dictates of restoring and promoting the financial sector. The response to the pandemic has clearly been very different even if warnings of the austerity to come (who is going to pay for the bloated state debt) have been commonplace, and the prospects for the developing world are particularly bleak. In this light, lessons from, or coming out of, the GFC may not have been learned but, hopefully, those from the pandemic may be stronger and more wide-ranging, especially if deriving from what was achieved within the crisis itself. In particular, the crash brought about by the pandemic prompted massive intervention to save lives and to govern, however much unevenly and successfully, economic and social reproduction by detailed intervention within each and across more or less every sector. The contrast can be drawn not only with the GFC but also with the, almost literally by comparison, slow-burning, environmental crisis which is a much greater (potential) source of loss of life and well-being and for which collective, progressive action is so vital.

Significantly, as already outlined in this chapter, the importance of the SoP approach for addressing environmental sustainability has been widely acknowledged. Hopefully, it will be able to inform future policymaking post-pandemic, alongside the provisioning systems, and associated cultures, of food, health, education, housing, transport, clothing and so much more. In a sense, this is the door against which the White House Conference on Food, Nutrition, and Health (Mande et al. 2020) is already pushing, however forcibly. It is worth quoting at length, to give a flavour of insights and responses to it, even at the expense of reinforcing points already made (p. 13):

> Poor diet is now the leading cause of poor health in the U.S., causing more than half a million deaths per year. The prevalence of obesity has risen sharply from 15% of adults and 5.5% of children in 1980 to 39.8% of adults and 18.5% of children in 2016. Nearly three in four (71.6%) American

adults are either overweight or have obesity. More than 100 million Americans—nearly half of all U.S. adults—suffer from diabetes or pre-diabetes, while one in three U.S. children born after 2000 is expected to develop Type 2 diabetes. Cardiovascular disease afflicts about 122 million people and causes roughly 840,000 deaths each year, with rates of coronary heart disease and obesity-related cancers increasing among younger adults. And, for the first time in American history, life expectancies falling, with declines for three consecutive years due in part to significant increases in midlife mortality from diet-related diseases

So much for the effects, what of the economic costs, p. 14:[28]

The economic costs of this new national nutrition crisis are staggering. Total U.S. healthcare expenditures have risen from 6.9% of gross domestic product (GDP) in 1970 to 17.9% in 2017. These rising medical costs, dominated by diet-related chronic health conditions, are crushing government budgets and private business growth. Total direct healthcare and indirect economic costs for cardiovascular diseases are estimated at $316 billion per year; for diabetes, at $327 billion per year; and for all obesity-related conditions, at $1.72 trillion per year. These amounts dwarf the annual budgets of many federal agencies, including the budgets of the Departments of Agriculture ($140 billion), Education ($72 billion), Homeland Security ($52 billion), and Justice ($28 billion), as well as the budgets of the National Institutes of Health ($39 billion), Centers for Disease Control and Prevention ($11 billion), Environmental Protection Agency ($5.7 billion), and Food and Drug Administration ($5.7 billion) … In summary, governments, businesses, farmers, and individuals all bear the health and environmental burdens of our food system—at massive expense. Globally, the externalities of our food system are estimated to total $12 trillion, an amount greater than the entire food sector's revenue.

In short, the Report, with reference to its externalities, can be seen to be situating the health system with the social determinants of health, with the food system to the fore—there, at least in part, but not in name, goes the SoP approach and its application through intersections of systems of systems, as discussed above in this closing chapter. Yet, the policy lessons do not seem to have been taken on board. For the very systems that transformed under- to over-nutrition within a generation or two are at most fiddled with to restore some sort of balance.

Five policy programmes are offered: to leverage the power of U.S. Department of Agriculture (USDA) programs; to utilize economic incentives; to protect children from harmful advertising and marketing; to equip health professionals with effective nutrition interventions and better nutrition knowledge; and to align agriculture, health, and sustainability, p. 7. These differ little from exactly the same proposals that have been around for decades, questioning the extent to which they will be implemented and/or effective if (fully) implemented—although the alignment of agriculture, health and sustainability is relatively novel in deference to unavoidable environmental crises. Indeed, it could be argued that the crises of obesity and the environment are the consequences of such policy framings if not the Report's substantive intent—essentially managing 'externalities' rather than pre-empting them. It is the food *system*, its material cultures and its intersections with other SoPs that need to be addressed!

In contrast, the SoP approach, then, has evolved to provide a comprehensive framework for guiding analysis into the drivers of consumption outcomes, not simply twiddling at the margins of market and the state. It offers a very broad framing of consumption, stretching across, and variously applied to, sectors from food to health services. The approach, then, provides a way to unpack the opaque workings of who gets what and how, as well as why. Equally significant is who is excluded and with what effects. The approach has been shown to be durable and versatile, with varieties of applications. Some studies have followed the approach in detail, examining the structures, processes, agents and relations which underpin provisioning as well as the material cultures that shape consumption decisions. Others have taken the SoP approach as one of a group of approaches on which to draw when considering consumption. Still others just use some core elements of the approach selectively, as for example, when the focus is on material cultures. As exemplified by both the grand scale of the GFC, the environment, the pandemic and the crisis of dietary diseases and the everyday access to health, water, housing and so on, the SoP approach offers an open, flexible if uncompromisingly grounded analytical framing for the challenges ahead in the worlds of scholarship, popular discourse, policymaking and activism.

Notes

1. See Cook, Crang & Thorpe (1998), and Warde and Martens (2000) and Shove (2003), respectively, for example. For an update on eating out, see Paddock, Warde & Whillans (2017).

2. This is not to suggest there was no dissent, especially from what might be termed the dualism police, charging the SoP approach in particular for invalid separations between one or other aspect, such as social and natural, material and cultural and so on. Indeed, Guthman (2002) parodies the SoP approach as such and for its supposed determinism attached to its political economy together with its corresponding neglect of reflexivity, not least because 'senses of taste … each maps onto any system of provision in different ways', p. 301/2. But, apart from at least implicitly and inconsistently accepting the SoP approach, this is exactly application and conclusion of the SoP approach itself. For discussion of responses to the dualisms critique of the SoP approach, see footnote in Chap. 5 and Fine (2002, 2004).
3. The Handbook is available from https://www.lancaster.ac.uk/fass/projects/esf/hndbkhome.htm.
4. See also van Vliet, Chappells & Shove (2005) and Southerton, Chappells & van Vliet (Eds.) (2004). Note that these contributions not only draw upon the SoP approach but anticipate much of what becomes stronger within the literature, as laid out below, especially in relation to focus on infrastructure, intersections across SoPs, and sustainability
5. Note that the third point on alternatives is indicative of an increasing politics and policy turn.
6. Note that two decades after its emergence, the SoP approach merits an entry by Evans in Southerton (Ed.) (2011) but alongside five or six hundred other entries. It does not even merit an index entry less than a decade later in Keller, Halkier, Wilska & Truninger (Eds.) (2017)! On the other hand, there is a whole chapter in the 2018 SAGE Handbook of Consumer Culture. But we are co-authors and it was included at our own suggestion (Bayliss, Fine & Robertson 2018b)!
7. It is interesting to speculate why this should be so, or not, by comparison with other topics and associated, more or less hegemonic approaches. Some do become drowned, even superseded, by those jumping on the bandwagon from other perspectives, as with social capital theory and the displacement of Bourdieu as its modern founder, Fine (2010). Others, such as the developmental state paradigm, remain relatively stable and narrow in their scope of approaches, and this is especially so of the welfare regime approach, see Fine and Pollen (2018) and Fine (2014), respectively. For topics such as globalization, neoliberalism and financialization, there are any number of approaches whatever the concepts' origins.
8. See also Tonkiss (2015) on moral economy in the context of infrastructure.
9. With reference to the SoP approach, Harris (2014, p. 7) sees the neoliberalization of social work in terms of marketisation, consumerization and managerialization, concluding that 'The extent to which these processes

have taken hold and the precise combinations in which they appear vary but they constitute a direction of travel in many countries.'

10. The role of SoP and other systems based approaches in relation to sustainability is discussed in more detail below.
11. See http://geography.exeter.ac.uk/research/groups/energypolicy/research/steppingup/ and also 'The Domestic Nexus: Interrogating the interlinked practices of water, energy and food consumption', https://www.sheffield.ac.uk/geography/nexus
12. For more on the iBuild project see https://research.ncl.ac.uk/ibuild/outputs.
13. It is no accident that there is a close relationship between the University of Leeds component of the iBuild programme and the Lili research.
14. For more on the Rebuilding Macro project see https://www.rebuilding-macroeconomics.ac.uk/trajectories-infranstructure.
15. As again and again in British rail, see Haines-Doran (2019).
16. For more on the Living Well Within Limits (Lili) project see https://lili.leeds.ac.uk/about/.
17. With system of provision mentioned many times without reference to its literature. Note that waste and recycling as topics have explicitly been drawn to SoP analyses on a number of occasions, not least OECD (2002).
18. See also Bell, Judson, Bulkeley, Powells, Capova & Lynch (2015) which also seems to take the SoP approach for granted.
19. See https://cordis.europa.eu/project/id/249430.
20. See https://www.ncbi.nlm.nih.gov/pmc/articles/PMC6477554/.

    The extent to which the gravity of the pandemic was accentuated by malnutrition among war-tired populations is unclear. However, the fact that the disease, even in serious forms, spread through countries that were neutral or completely uninvolved in the war, such as Spain, seems to suggest that malnutrition was not a key factor.

21. This comes from the foreword to the report. Within its body, p. 12/13:

    Fifty years after the original White House Conference on Food, Nutrition, and Health, the U.S. faces a very different national nutrition crisis, together with newer challenges of climate change and sustainability. While calorie malnutrition in America has been largely eradicated, changes to our food system accompanied by persistent poverty and increasing economic inequality have created a crisis of diet-related obesity, diabetes, and other chronic diseases, and widened other disparities in the accessibility and affordability of nutritious foods. Not addressed at the 1969 conference but highly relevant today is the added challenge of feeding a growing population healthfully and sustainably and without exacerbating climate change and depleting natural resources.

22. Other leading preventable causes of not being able to join the military include lacking adequate education and having a history of crime or drug use (Maxey, Bishop-Josef & Goodman 2018).
23. This astonishing report by Maxey et al. (2018), is preceded by the even more astonishing (Still) too Fat to Fight, Mission: Readiness (2010, 2012).
24. See https://www.foodnavigator.com/Article/2020/04/22/Coronavirus-and-obesity-Doctors-take-aim-at-food-industry-over-poor-diets.
25. See https://www.foodnavigator.com/Article/2020/04/22/Coronavirus-and-obesity-Doctors-take-aim-at-food-industry-over-poor-diets.
26. Interestingly, height has become a favoured index of under-nutrition for times past; today, as a sign of changed times, it tends to have been displaced by the body-mass index, BMI, a measure of weight relative to height!
27. For a telling account of the rush to publish, from academia to social media, and for its consequences, see https://www.theatlantic.com/health/archive/2020/04/pandemic-confusing-uncertainty/610819/.
28. Indeed:

    Healthcare costs are imperiling other priorities in federal and state budgets and in the economy as a whole. Healthcare spending rose from 5% of all federal spending in 1970 to 28% in 2018 and from 11.3% of state budgets in 1989 to 28.7% in 2016. For U.S. businesses, healthcare expenditures rose (in constant 2017 dollars) from $79 billion in 1970 to $1.18 trillion in 2017). On a per capita basis, annual healthcare spending has increased from $1797 per person in 1970 to $10,739 per person in 2017 (in constant 2017 dollars).

References

Bayliss, K., Fine, B., & Robertson, M. (Eds.). (2018a). Material cultures of financialisation. London: Routledge, reproduced special issue of *New Political Economy*, 2017, 22(4), 335–370.

Bayliss, K., Fine, B., & Robertson, M. (2018b). The systems of provision approach to understanding consumption. In O. Kravets, P. Maclaran, S. Miles, & A. Venkatesh (Eds.), *The SAGE handbook of consumer culture* (pp. 27–42). London: Sage.

Bayliss, K., Fine, B., Robertson, M., & Saad Filho, A. (2020). *Neoliberalism, financialisation and welfare: The political economy of social provision in the UK*. Cheltenham: Edward Elgar, in preparation.

Bayliss, K., & Mattioli, G. (2018). Privatisation, inequality and poverty in the UK: Briefing prepared for UN Rapporteur on extreme poverty and human rights. Sustainability Research Institute Working Paper Series, No. 116. Retrieved from https://www.see.leeds.ac.uk/fileadmin/Documents/research/sri/workingpapers/SRIPs-116.pdf.

Bell, S., Judson, E., Bulkeley, H., Powells, G., Capova, K., & Lynch, D. (2015). Sociality and electricity in the United Kingdom: The influence of household dynamics on everyday consumption. *Energy Research & Social Science, 9*, 98–106.

Bonizzi, B., & Churchill, J. (2017). Pension funds and financialisation in the European Union: Fondos de pensiones y financiarizacion en la Union Europea. *Revista de Economia Munidal, 46*, 71–90.

Brooks, A. (2015a). *Clothing poverty: The hidden world of fast fashion and second-hand clothes.* London: Zed Books.

Brooks, A. (2015b). Systems of provision: Fast fashion and jeans. *Geoforum, 63*, 36–39.

Brown, A., & Robertson, M. (2014). *Economic evaluation of systems of infrastructure provision: Concepts, approaches, methods.* IBUILD/Leeds Report. University of Leeds, Leeds. Retrieved from https://research.ncl.ac.uk/ibuild/outputs/reports/9940_iBuild_report_print_version%20WEB.pdf.

Bulkeley, H., & Askins, K. (2009). Waste interfaces: Biodegradable waste, municipal policy and everyday practice. *The Geographical Journal, 175*(4), 251–260.

Case, A., & Deaton, A. (2020). *Deaths of despair and the future of capitalism.* Princeton: Princeton University Press.

Chappells, H., & Shove, E. (2000). Organising energy: Consumption, production, and co-provision. In *American council for an energy-efficient economy 2000 summer study on energy efficiency in buildings* (pp. 39–50). Washington, DC: American Council for and Energy-Efficient Economy. Retrieved from https://www.aceee.org/files/proceedings/2000/data/papers/SS00_Panel8_Paper04.pdf.

Churchill, J. (2013). *Towards a framework for understanding the recent evolution of pension systems in the European Union.* FESSUD Working Paper Series, No. 12. Retrieved from http://fessud.eu/wp-content/uploads/2013/04/Towards-a-framework-for-understanding-the-recent-evolution-of-pension-systems-in-the-European-Union-FESSUD-working-paper-12.pdf.

Churchill, J. (2018). Financialisation of pensions in Europe: Gendered consequences. Mimeo.

Churchill, J. (2019). *Evolving financial institutions and their macroeconomic implications: Case of Poland.* Mimeo, paper presented at EAEPE Conference, Warsaw September 2019.

Cook, I., Crang, P., & Thorpe, M. (1998). Biographies and geographies: Consumer understandings of the origins of foods. *British Food Journal, 100*(3), 162–167. https://doi.org/10.1108/00070709810207522.

De Feijter, F., Van Vliet, B., & Spaargaren, G. (2019). Energy efficient housing through organized interactions? Conceptualizing the roles of householders and providers in housing retrofitting in the Netherlands and China. *Housing, Theory and Society*. https://doi.org/10.1080/14036096.2019.1658623.

Evans, D. (2011). Systems of provision. In D. Southerton (Ed.), *Encyclopedia of consumer culture* (p. 1425). London: Sage.

Fine, B. (2002). *The world of consumption*. London: Routledge.

Fine, B. (2004). Debating production-consumption linkages in food studies. *Sociologia Ruralis, 44*(3), 332–342.

Fine, B. (2005). Addressing the consumer. In F. Trentmann (Ed.), *The making of the consumer: Knowledge, power and identity in the modern world* (pp. 291–311). Oxford: Berg.

Fine, B. (2010). *Theories of social capital: Researchers behaving badly*. London: Pluto.

Fine, B. (2013a). Economics: Unfit for purpose. *Review of Social Economy, LXXI* (3), 373–389. Shorter version of "Economics—Unfit for purpose: The Director's Cut", SOAS Department of Economics Working Paper Series, No. 176, 2013. Retrieved from https://www.soas.ac.uk/economics/research/workingpapers/file81476.pdf.

Fine, B. (2013b). Consumption matters. *Ephemera, 13*(2), 217–248.

Fine, B. (2014). *The continuing enigmas of social policy*. United Nations Research Institute for Social Development Working Paper Series, No. 2014-10. Retrieved from http://www.unrisd.org/unrisd/website/document.nsf/(httpPublications)/30B153EE73F52ABFC1257D0200420A61?OpenDocument.

Fine, B. (2020a). Framing social reproduction in the age of financialisation. In A. Santos & N. Teles (Eds.), *Financialisation in the European periphery: Work and social reproduction in Portugal* (pp. 257–272). London: Routledge.

Fine, B. (2020b). Situating PPPs. In J. Gideon & E. Unterhalter (Eds.). *Critical reflections on public private partnerships*. London: Routledge, forthcoming.

Fine, B., & Pollen, G. (2018). The developmental state paradigm in the age of financialization. In M. Hyland & R. Munck (Eds.), *Handbook on development and social change* (pp. 211–227). Cheltenham: Edward Elgar.

Foden, M., Browne, A., Evans, D., Sharp, L., & Watson, M. (2019). The water–energy–food nexus at home: New opportunities for policy interventions in household sustainability. *The Geographical Journal, 185*, 406–418.

Foster, C., McMeekin, A., & Mylan, J. (2012). The entanglement of consumer expectations and (eco) innovation sequences: The case of orange juice. *Technology Analysis & Strategic Management, 24*(4), 391–405.

Goodman, D., & DuPuis, E. (2002). Knowing food and growing food: Beyond production-consumption debate in the sociology of agriculture. *Sociologia Ruralis, 42*(1), 5–22.

Guthman, J. (2002). Commodified meanings, meaningful commodities: Re-thinking production–consumption links through the organic system of provision. *Sociologia Ruralis, 42*(4), 295–311.

Haines-Doran, T. (2019). *Understanding the changing role of central government in Britain's railways since privatisation: A systems of provision approach.* Unpublished PhD thesis, School of Oriental and African Studies, University of London.

Hansen, A. (2018a). Meat consumption and capitalist development: The meatification of food provision and practice in Vietnam. *Geoforum, 93*, 57–68.

Hansen, A. (2018b). Consuming development: Capitalism, economic growth and everyday life. *Blog Entry.* Retrieved from https://developingeconomics.org/2018/01/11/consuming-development-capitalism-economic-growth-and-everyday-life/.

Harris, J. (2014). (Against) Neoliberal social work. *Critical and Radical Social Work, 2*(1), 7–22.

Iacovidou, E., Millward-Hopkins, J., Busch, J., Purnell, P., Velis, C., Hahladakis, J., Zwirner, O., & Brown, A. (2017). A pathway to circular economy: Developing a conceptual framework for complex value assessment of resources recovered from waste. *Journal of Cleaner Production, 168*, 1279–1288.

Jackson, P., Russell, P., & Ward, N. (2004). *Commodity chains and the politics of food.* Cultures of Consumption Working Paper Series, No 18. Retrieved from http://www.consume.bbk.ac.uk/publications.html.

Judson, E., Bell, S., Bulkeley, H., Powells, G., & Lyon, S. (2015). The co-construction of energy provision and everyday practice: Integrating heat pumps in social housing in England. *Science & Technology Studies, 28*(3), 26–53.

Kantamaturapoj, K., Oosterveer, P., & Spaargaren, G. (2012). Emerging market for sustainable food in Bangkok. *International Journal of Development and Sustainability, 1*(2), 268–279.

Keller, M., Halkier, B., Wilska, T.-A., & Truninger, M. (Eds.). (2017). *Routledge handbook on consumption.* London: Routledge.

Mande, J., Willett, W., Auerbach, J., Bleich, S., Broad Leib, E., Economos, C., Griffin, T., Grumbly, T., Hu, F., Koh, H., Mozaffarian, D., Pérez-Escamilla, R., Seligman, H., Story, M., Wilde, P., & Woteki, C. (2020). Report of the 50th Anniversary of the White House Conference on Food, Nutrition, and Health: Honoring the Past, Taking Actions for our Future. Boston, MA.

Mansvelt, J. (2005). *Geographies of consumption.* London: Sage.

Mattioli, G., Roberts, C., Steinberger, J., & Brown, A. (2020). The political economy of car dependence: A systems of provision approach. *Energy Research & Social Science, 66.* https://doi.org/10.1016/j.erss.2020.101486.

Maxey, H., Bishop-Josef, S., & Goodman, B. (2018). *Unhealthy and unprepared: National security depends on promoting healthy lifestyles from an early age*. USA: Council for a Strong America. Retrieved from https://strongnation.s3.amazonaws.com/documents/484/389765e0-2500-49a2-9a67-5c4a090a215b.pdf.

Meadows, D. (2008). *Thinking in systems: A primer*. London: Earthscan.

Mission: Readiness. (2010). Too fat to fight: Retired military leaders want junk food out of America's schools. Retrieved from https://strongnation.s3.amazonaws.com/documents/24/ec1aefe5-f170-4f75-ade5-0e8063430f16.pdf?1469638848&inline;%20filename=%22Too%20Fat%20to%20Fight.pdf%22.

Mission: Readiness. (2012). Still too fat to fight. Retrieved from https://mission-readiness.s3.amazonaws.com/wp-content/uploads/Still-Too-Fat-To-Fight-Report.pdf.

Monahan, J., & Powell, J. (2011). A comparison of the energy and carbon implications of new systems of energy provision in new build housing in the UK. *Energy Policy, 39*(1), 290–298.

Moore, T., & Dalton, T. (2016). Structuring housing provision for urban sustainability. In R. Horne, J. Fien, B. Beza, & A. Nelson (Eds.), *Sustainability citizenship in cities* (pp. 80–92). London: Routledge.

OECD. (2002). *Sector case studies series household energy & water consumption and waste generation: Trends, environmental impacts and policy responses*. Report by Working Party on National Environmental Policy. Paris: Organisation for Economic Cooperation and Development. Retrieved from http://www.oecd.org/officialdocuments/publicdisplaydocumentpdf/?doclanguage=en&cote=env/epoc/wpnep(2001)15/final.

Paddock, J., Warde, A., & Whillans, J. (2017). The changing meaning of eating out in three English cities 1995–2015. *Appetite, 119*, 5–13.

Pirgmaier, E., & Steinberger, J. (2019). Roots, riots, and radical change—A road less travelled for ecological economics. *Sustainability, 11*(7). https://doi.org/10.3390/su11072001.

Reisch, L., Cohen, M., Thøgersen, J., & Tukker, A. (2016). Frontiers in sustainable consumption research. *GAIA—Ecological Perspectives for Science and Society, 24*(4), 234–240.

Rinkinen, J. (2013). Electricity blackouts and hybrid systems of provision: Users and the 'reflective practice'. *Energy, Sustainability and Society, 3*(25). https://doi.org/10.1186/2192-0567-3-25.

Rinkinen, J., Shove, E., & Smits, M. (2019). Cold chains in Hanoi and Bangkok: Changing systems of provision and practice. *Journal of Consumer Culture, 19*(3), 379–397.

Saritas, S. (2014). *Review of pension provision across the European Union countries*. FESSUD Working Paper Series, No 13. Retrieved from http://fessud.eu/wp-content/uploads/2013/04/REVIEW-OF-THE-PENSION-PROVISION-ACROSS-THE-EUROPEAN-UNION-COUNTRIES_13.pdf.

Saritas, S. (2016). *Financialization and Turkish pension reform*. Unpublished PhD thesis, School of Oriental and African Studies, University of London.

Saritas, S. (2020). Financialisation of pensions: The case of Turkey. *Global Social Policy, 20*(1), 60–79.

Schafran, A., Noah Smith, M., & Hall, S. (2020). *The spatial contract: A new politics of provision for an urbanised planet*. Manchester: Manchester University Press.

Serrano, A., & Brooks, A. (2019). Who is left behind in global food systems? Local farmers failed by Colombia's avocado boom. *Environment and Planning E: Nature and Space, 2*(2), 348–367.

Seyfang, G. (2007). Cultivating carrots and community: Local organic food and sustainable consumption. *Environmental Values, 16*(1), 105–123.

Shove, E. (2003). *Comfort, cleanliness and convenience: The social organization of normality*. Oxford: Berg.

Shove, E. (2015). Matters of practice. In A. Hui, T. Schatzki, & E. Shove (Eds.), *The nexus of practices: Connections, constellations, practitioners* (pp. 155–168). London: Routledge.

Shove, E., & Walker, G. (2010). Governing transitions in the sustainability of everyday life. *Research Policy, 39*(4), 471–476.

Southerton, D., Chappells, H., & van Vliet, B. (Eds.). (2004). *Sustainable consumption: The implications of changing infrastructures of provision*. Cheltenham: Edward Elgar.

Tonkiss, F. (2015). Afterword: Economies of infrastructure. *City, 19*(2–3), 384–391.

van der Walt, K. (2016). *Thorough-bread? A study of the system of provision of bread in South Africa*. Masters Research Report, Johannesburg, University of the Witwatersrand. Retrieved from https://pdfs.semanticscholar.org/692b/caf67825d8cfc7bb42a7b07bd7b458bda2dc.pdf.

van Vliet, B., Chappells, H., & Shove, E. (2005). *Infrastructures of consumption environmental innovation in the utility industries*. London: Routledge.

Warde, A. (2016). *The practice of eating*. Cambridge: Polity Press.

Warde, A., & Martens, L. (2000). *Eating out: Social differentiation, consumption and pleasure*. Cambridge: Cambridge University Press.

Watts, D., Ilbery, B., & Maye, D. (2005). Making reconnections in agro-food geography: Alternative systems of food provision. *Progress in Human Geography, 29*(1), 22–40.

Werna, E. (1998). Urban management, the provision of public services and intra-urban differentials in Nairobi. *Habitat International, 22*(1), 15–26.

Wheeler, K. (2018). The moral economy of ready-made food. *The British Journal of Sociology, 69*(4), 1271–1292.

Williams, D., Spotswood, F., Parkhurst, G., & Chatterton, T. (2019). Practice ecology of sustainable travel: The importance of institutional policy-making processes beyond the traveller. *Transportation Research Part F, 62*, 740–756.

Winter, J. (1980). Military fitness and civilian health in Britain during the First World War. *Journal of Contemporary History, 15*(2), 211–244.

Author Index[1]

[1] Note: Page numbers followed by 'n' refer to notes.

K. Bayliss, B. Fine, *A Guide to the Systems of Provision Approach*,
https://doi.org/10.1007/978-3-030-54143-9

Subject Index[1]

[1] Note: Page numbers followed by 'n' refer to notes.

K. Bayliss, B. Fine, *A Guide to the Systems of Provision Approach*, https://doi.org/10.1007/978-3-030-54143-9

D

Q

R

T

The manufacturer's authorised representative in the EU is Springer Nature Customer Service Centre GmbH, Europaplatz 3, 69115 Heidelberg, Germany. If you have any concerns regarding our products, please contact ProductSafety@springernature.com

Printed and bound by CPI Group (UK) Ltd, Croydon, CR0 4YY

30/09/2026

02225325-0002